LOSING PARADISE

Voices in Development Management

Series Editor:
Margaret Grieco
Napier University, Scotland

The Voices in Development Management series provides a forum in which grass roots organisations and development practitioners can voice their views and present their perspectives along with the conventional development experts. Many of the volumes in the series will contain explicit debates between various voices in development and permit the suite of neglected development issues such as gender and transport or the microcredit needs of low income communities to receive appropriate public and professional attention.

Also in the series

Tourism, Development and Terrorism in Bali
Michael Hitchcock and I Nyoman Darma Putra
ISBN 978 0 7546 4866 6

Women Miners in Developing Countries
Pit Women and Others
Edited by Kuntala Lahiri-Dutt and Martha Macintyre
ISBN 978 0 7546 4650 1

Africa's Development in the Twenty-first Century
Pertinent Socio-Economic and Development Issues
Edited by Kwadwo Konadu-Agyemang and Kwamina Panford
ISBN 978 0 7546 4478 1

Accountability of the International Monetary Fund
Edited by Barry Carin and Angela Wood
ISBN 978 0 7546 4523 8

Social Exclusion and the Remaking of Social Networks
Robert Strathdee
ISBN 978 0 7546 3815 5

Design and Determination
The Role of Information Technology in Redressing
Regional Inequities in the Development Process
Stephen E. Little
ISBN 978 0 7546 1099 1

Losing Paradise
The Water Crisis in the Mediterranean

GAIL HOLST-WARHAFT and TAMMO STEENHUIS
Cornell University, USA

Routledge
Taylor & Francis Group
LONDON AND NEW YORK

First published 2010 by Ashgate Publishing

Published 2016 by Routledge
2 Park Square, Milton Park, Abingdon, Oxon OX14 4RN
711 Third Avenue, New York, NY 10017, USA

Routledge is an imprint of the Taylor & Francis Group, an informa business

British Library Cataloguing in Publication Data
Losing paradise : the water crisis in the Mediterranean. --
(Voices in development management)
1. Water-supply--Middle East. 2. Water resources
development--Middle East.
I. Series II. Holst-Warhaft, Gail, 1941- III. Steenhuis,
Tammo S.
333.9'1'00956-dc22

Library of Congress Cataloging-in-Publication Data
Losing paradise : the water crisis in the Mediterranean / [edited] by Gail Holst-Warhaft and Tammo Steenhuis.
p. cm. -- (Voices in development management)
Includes bibliographical references and index.
ISBN 978-0-7546-7573-0 (hardback)
1. Water-supply--Mediterranean Region--Management. 2. Water resources development--Mediterranean Region. I. Holst-Warhaft, Gail, 1941- II. Steenhuis, Tammo S.
HD1697.5.M43L67 2009
333.91009182'2--dc22

2009033304

ISBN 9780754675730 (hbk)

Contents

List of Figures

List of Tables

Notes on Contributors

Symeon Christodoulou holds a PhD in Civil Engineering from Columbia University and BSc, MSc, and MPhil degrees from the same university. He joined the Brooklyn Polytechnic University as an Assistant Professor and head of the Construction Management Program (1998–2003). In 2004 he joined the University of Cyprus. Dr. Christodoulou is the author of several scientific publications. In 2008 he was elected as Cyprus's national representative to EU's Cost Management Committee on Transport and Urban Development (TUD). Dr Christodoulou's main research interests are in the areas of risk assessment and rehabilitation of urban infrastructure, fully integrated and automated project processes, information technology, soft-computing and artificial agents.

Eriberto Eulisse is an anthropologist and art curator. He is director of the Water Civilization International Centre, a non-profit organization based in Venice, Italy (www.civiltacqua.org). He graduated in Social Sciences from the University of Bologna and specialized in Cultural Anthropology at the Sainsbury Center for Visual Arts, Norwich (UK). He has worked as curator of the Ethnographic Museum of Trento, in the Italian Alps, and as project officer of the European Commission in Brussels (Marie Curie Programme). Also he worked as associate curator of the exhibition *Authentic/Ex-centric: Africa In and Out of Africa*, the first Pavilion of African Art at the Venice Biennale, organized in 2001 by the Forum for African Arts (New York). His recent contributions on water issues include: *Water, Culture, Society: Managing Water Resources in European Mountain Environments* (ed., 2008) and *Silis, Annali di Civiltà dell'Acqua* (ed., 2008).

Nadim Farajalla Before joining the American University of Beirut, Farajalla was a Senior Environmental Engineer at Dar Al-Handasah Consultants (Shair and Partners) and prior to that Senior Scientist at Stone Environmental Inc. (Montpelier, VT, USA). His experience includes extensive water resource and storm-water drainage design projects and the development of master plans for infrastructure development and environmental appraisal in countries such as Lebanon, Turkey, Angola, Nigeria, Morocco, Saudi Arabia and the USA. His current research efforts have focused on local and regional water resources needs and problems, with emphasis on integrated water resource management and issues related to climate change.

Gail Holst-Warhaft directs the Mediterranean Studies Initiative at Cornell University's Institute for European Studies, and is a Faculty Fellow of the

Cornell Center for a Sustainable Future, and is an adjunct professor in the departments of Classics, Comparative Literature, and Near Eastern Studies. She is also a poet, musician, and translator. She has written widely about the Mediterranean, particularly Greek culture. Her books include *Road to Rembetika* (1975), *Theodorakis: Myth and Politics in Modern Greek Music* (1980), *Dangerous Voices: Women's Laments and Greek Literature* (1992), *The Cue for Passion: Grief and its Political Uses* (2000) and *Penelope's Confession* (2007). She co-teaches a course on water in the Mediterranean at Cornell University with Keith Porter and Tammo Steenhuis.

Nicholas S. Hopkins is Professor of Anthropology Emeritus at the American University in Cairo, where he was a regular faculty member from 1975 to 2006. In retirement he continues to reside in Egypt. He has conducted and supervised various research projects in rural Egypt on irrigation and on environmental issues. In 2001 he and his colleagues published *People and Pollution: Cultural constructions and Social Action in Egypt* (American University in Cairo Press), comparing various sites in rural and urban Egypt. He has also carried out research in Mali, Tunisia, and India. In all these research projects he has stressed the impact of broader development plans on the lives and life chances of the poor majority.

Gaspar Mairal is Associate Professor of Anthropology at the University of Saragossa in Spain. He has widely researched the social impact of dam construction in the Spanish Pyrenees, the cultural background of water policies in the 20th century and most recently, the study of the communications of risk. At present he leads a project researching the history and the narratives of risk. He is the author of several books and articles on water including *Agua, Tierra, Riesgo y Supervivencia* (1997), *Los aragoneses y el agua* (2000), "From Economism to Culturalism: the Social and Cultural Construction of risk in the river Esera" (1998) and "The Invention of a Minority: A case study from the Aragonese."

Zeyad Makhamreh holds an Assistant professorship in the Department of Geography at the University of Jordan. He holds a BSc in Soil and Irrigation Science, and a Master's Degree in land use and natural resource management. His doctoral degree was in applied remote sensing in land degradation assessment and land use change in Jordan. He has a long experience in the fields of natural resources management and sustainable agricultural practices in Jordan and has acquired extensive expertise in various aspects of applied environmental science. His research focus is innovative approaches in remote sensing analysis as applied to watershed management, and monitoring of land use changes using earth observation data. He has many publications in international journals in the field of land use, watershed management and remote sensing image analysis.

Keith S. Porter was the Director of the New York State Water Resources Institute from 1985 to 2008. In this capacity, he was the Principal Investigator for multiple inter-disciplinary research projects with the objective of assisting government agencies, and communities, including the agricultural community, address their water resource problems and needs. Since 2005 he has been an adjunct Professor of Law at Cornell University Law School where he is also a Faculty Affiliate in the Cornell American Indian Program. Keith has served on multiple advisory committees and boards and is currently the Chairman of the new Water Discovery Center in the Catskills, New York.

Tammo S. Steenhuis received his Kandidaats (BS) degree and the Ingenieur (MS) degree in Land and Water Management of the tropics from Wageningen Agricultural University, Wageningen, (Netherlands) and MS and PhD degrees from the Agricultural Engineering Department at the University of Wisconsin, Madison. Since 1978 he has been a faculty member in the Department of Biological and Environmental Engineering at Cornell University where he is an International Professor of Water Management. He is currently directing the Cornell Masters program in Watershed Management and Hydrology at the University of Bahir Dar in Ethiopia and co-teaches a course on the right to water with Gail Holst-Warhaft and Keith Porter. His research encompasses the movement of colloids at the pore scale and large scale watershed studies of the distributed runoff response of watersheds such as New York City source watersheds in the Catskill Mountains and the Blue Nile in Ethiopia. He is a recipient of the Darcy Medal of the European Geosciences Union and a fellow of the American Geophysical Union.

George Stournaras is Professor of Hydrology and Engineering Geology at the University of Athens. He received his doctoral degree in Applied Geology from the University of Grenoble and has published more than 100 papers in his field. He has also published three books, including *Water: Environmental Nature and Course* which won the Academy of Athens Award in 2007. He is a member of many international scientific associations including the International Association of Environmental Hydrologists, and the International Water History Association.

Preface and Acknowledgements

The editors and principal authors of this book work in fields that rarely overlap. Despite the differences in their approach to water problems, the editors, together with Professor Keith Porter (Cornell Law School) have cooperated to develop and teach a graduate course at Cornell University about water management, culture, and law in the Mediterranean Basin. Our collaboration is the rationale for this volume. During our discussions with senior water researchers at Cornell we all reached the same conclusion: no-one understands the need for an integrated approach to the crisis of water better than those who have worked in the developing regions of the world. The current problems of water will not be solved by technology, law or politics, and yet all these must play a role if we are to overcome the crisis. People's attitudes to water, its use and abuse are culturally specific and directly related to the stage of development reached by their society. The problem of water is always linked to economic development, and the southern and eastern shores of the Mediterranean are disproportionately affected by the crisis. The water problems facing the inhabitants of Italy or Greece may seem minor by comparison, but the northern and southern shores of the Mediterranean are becoming increasingly interdependent as the sea that links them deteriorates and the population surrounding it grows.

The population of the Mediterranean states has doubled in the last forty years and is expected to reach over 600 million in 2050. Tourists will add at least another 350 million to that figure in summer. Twenty million people in the south-eastern Mediterranean have no access to drinking water and only 55 per cent of coastal cities with more than 10,000 inhabitants have sewage treatment facilities (World Bank: 2008). The effects of climate change exacerbate the problem of water scarcity and pollution. Scholars may debate the legitimacy of considering the Mediterranean as a historical or cultural unity, but they cannot afford to deny the common problems of its coastal states as they confront the present crisis in fresh water resources. The problem is not only an environmental one. A growing number of scholars and researchers worldwide are concerned about the potential for conflicts over scarce water in the region.

In the summer of 2008, the editors invited a group of scholars from the Mediterranean and the US to a series of meetings in areas of severe water stress in Greece. In the town of Nafplion we held a day-long conference attended by mayors from all the municipalities of the Argive Plain, by Greek water specialists and members of the public.

Like many areas of the Mediterranean, the plain of Argos has become a center for the intensive cultivation of a single valuable crop (oranges) requiring large amounts of water. Tens of thousands of wells have been dug to satisfy the farmers' needs, and millions of tons of pesticides and fertilizers have been dumped on the fields. The result is that the local water is unsafe to drink and salt has infiltrated the ground water. No-one at the meeting, from politicians to farmers, disputed the facts, nor were they able to offer any feasible solutions. Water is traditionally supplied free to farmers, and as long as northern Europeans want fresh oranges in their supermarkets, and politicians depend on the agricultural sector's votes, there is very little incentive to change the situation. When the conference ended, a well-known Greek singer performed songs about water with local schoolchildren. Hundreds of townspeople sat in the open square listening. Perhaps their children will be the ones to change the situation, or perhaps it will take a courageous mayor or local leader to put a stop to illegal drilling or ban the use of poisonous fertilizers.

The most optimistic sign we found on our trip was in eastern Crete where desertification is a serious problem. In the town of Neapolis the local mayor was waging an energetic campaign in the schools and his community to improve the environment and encourage local farmers to conserve water. We realized that the mayor of a small town could achieve more than the EU's Water Framework or the national government. He was a four-term mayor who knew his constituency, and understood how to communicate with it. Perhaps it is at this level that changes in behavior towards water can best be achieved, but the problem is complex and demands innovative, multi-dimensional solutions.

Together with the authors who have contributed to this volume, we are committed to an ongoing dialogue with scholars, local officials, students, artists and schoolchildren about water in the Mediterranean, and we would like to express our thanks to all of those who have helped us think about ways to solve the crisis. At Cornell, we are especially grateful to two of the "grand old men" of water management, Professors Gil Levine and Peter Loucks, who have shared the wisdom of their many years of experience with us and taught in our course. Keith Porter has been our loyal fellow conspirator in the efforts to form a Mediterranean water team and our co-teacher. We also thank the staff of the Cornell Institute for European Studies, especially Catherine Perkins, and Sydney Van Morgan for generous assistance, and to student assistants Jake Lee and Christine Chung. We are grateful for the generous support of the Cornell Center for a Sustainable Future and to the Director and Fellows of the Society for the Humanities who chose "Water" as their topic for 2008–9, and helped us think about it in creative ways. In Greece we thank Panayiotis Nektarios, of the Agricultural University of Athens for his astounding organization, Nikos Kastrinakis, Mayor of Neapolis, a lonely crusader for conserving water, Katerina Polymerou Kamilaki, director of the Folklore Museum of Athens, and the singer and children's choir director Mariza Koch, whose songs reminded her audience of the precious nature of water. We also thank the scholars who joined us on our travels: Ellen Graber, Eriberto Eulisse, John Nieber, Rony Wallach and Zeynep Zaimoglu, and our fellow-authors and collaborators in this volume. Finally, I would like to thank Margaret Grieco, Professor of Transport and Society, Napier University, Edinburgh, who suggested Ashgate Publishing to us and encouraged the publication of this volume and Maude Barlow, tireless crusader for the human right to water.

Gail Holst-Warhaft and Tammo Steenhuis

PART I
The Cultural Background

Chapter 1
Losing Paradise: The Water Crisis in the Mediterranean

Gail Holst-Warhaft

> Life is like a glass of cold water. ... and have you quenched your thirst?
>
> Old Cretan proverb.

> Water dreaming (Ngapa) is one of the great dreamings of central Australia and a key generative principle crucial to the life cycle … the role of the water men or women, for any community, is vital: their task is to sing in the rains.
>
> Notes for an exhibition of Australian Aboriginal paintings.

Introduction

In this chapter I would like to suggest that not only is it impossible to address the problems of water in the Mediterranean Basin without an understanding of the cultural factors that affect the way people in the region use water, but that culture, in its broadest and narrowest sense, including its historical and contemporary manifestations can and should be enlisted in the effort to redress the current crisis. It is misleading to consider water only as a form of 'natural capital.' Wherever water has been a scarce resource, it has a cultural importance that transcends the economic implications of the term 'capital.' Because of its traditional associations with the rituals of daily life, religious and secular, and with the human imagination, water has a significance that is both local and transnational. The Mediterranean itself, it can be argued, has always been an imagined space. Not only around its shores, but beyond them, people have dreamed of the Mediterranean area as an ideal *topos*. This idealization of the region can be dismissed as romantic and delusional, but it has provoked men to go to war to defend it and inspired the work of generations of artists. The power of the Mediterranean to inspire can and should be harnessed to defend the degradation and exploitation of perhaps its most precious resource: water.

Mediterranean Waters

> 'Άριστον μὲν 'ύδωρ

Pindar's first Olympian ode begins with the words 'most excellent is water.' After water, the poet praises gold and only then the Olympian hero Hieron, winner of the horse race in the games of 476 BC. Beliefs about water are deeply embedded in all the cultures of the Mediterranean, and reflected in their ethics, rituals, and religions. The belief that all life began with water was common among the ancient societies of the Mediterranean. The Egyptian hieroglyph for water was a wavy line that resembled a snake. Carried into the Semitic languages it became the letter *mem*, representing mayim (water) and in Latin, the letter M (Hillel, 1994: 23). In Greek mythology, Okeanos was a stream born of Earth (Ge) and Sky (Uranus). Okeanos's 3,000 sons were rivers, and his 3,000 daughters, the Oceanids. An endless series of descendents of these early water deities include Zeus, bringer of the rains, and Poseidon, who, despite his familiar association with the sea, was also a fresh water deity responsible for floods.

Like the sea, fresh water was unpredictable; in much of the Mediterranean, floods were both beneficial and destructive. The myth of Isis and Osiris, recorded in the pyramids of Saqqara in 2500 BC, accentuates the central role of the annual floods in Egyptian life. Osiris and his sister/wife Isis were credited with bringing order and civilization to Egypt, but Osiris's evil brother Set murdered Osiris and dismembered his body. Weeping bitterly, Isis set out to find her husband's body. The sun god Ra heard her lament and helped her in her search. Together, they found most of the pieces of the dead Osiris whom Isis was able to restore to life. The story of Osiris's death and resurrection were linked, in Egyptian belief, to the annual cycle of drought and flood.

When Herodotus visited Egypt, in the mid-5th century BC, he likened the scene in peak flood season to a country converted to a sea … 'the towns look like the islands of the Aegean'(*Histories*). Without these floods, the country could not have existed (*aut Nilus, aut nihil*, the Romans observed). To the Egyptians themselves, the floods were as mysterious as the origins of the Nile itself and were associated with the fat god Hapi. Each year the representative of the gods on earth, the ruling Pharaoh, would throw a written order into the river, commanding it to rise (de Châtal, 2007: 66).

The annual flooding of the Nile was so crucial to the survival of Egypt that its height was measured carefully by 'Nilometers' installed at Aswan, Karnak and Roda Island. A 'Nile Crier' was employed to announce the measurements taken at these locations in the streets of Cairo (de Châtal, 2007: 67).[1] The annual beneficence of the Nile floods continued to be celebrated in Egypt until the 20th century. Each year a dam was built of earth and a small mound called a 'bride' set in front of it with grain sown on top. The washing away of the 'bride' by the male river ensured fertility. In autumn an effigy of Osiris was molded from a mixture of earth and corn, symbolizing the resurrection of the land. The construction of the Aswan Dam ended the cycle of annual flood and unpredictable renewal. The

1 The optimum height for the flood was reckoned at 16 cubits, and in the Greco-Roman period, a Nile cult developed associated with the number 16. In the 1st century AD, the Emperor Trajan had a medal struck that depicted the Nile as a male god. A winged figure on the medal pointed to the Roman numeral XVI (de Châtal, 2007: 67).

greatest of Mediterranean rivers was tamed, floods could be controlled, and some of Egypt's water problems appeared to have been solved, but the water from the Nile is no longer sufficient to supply the demands of a rapidly-growing population, 96 per cent of whom live in the Nile Valley.

Water is equally important in the cosmogony of the Jewish bible, where the spirit of god hovers over 'the face of the deep' (Genesis II). When God first created heaven and earth, nothing grew 'for the lord God had not caused it to rain upon the earth' (Genesis 2: 5). When the rain came it created Paradise, a garden where life could flourish because of water. The banishing of Adam and Eve is also a return to the desert lands that preceded the rains. Whether or not it was caused by a change in the climate of Egypt, the Exodus was probably a historical migration through the desert of Sinai towards a more fertile 'Promised Land.' Whatever the case, the Biblical migration begins with an extraordinary water event: the 'parting of the waters' and the drying up of the River Jordan.

Another dramatic water event, the flood that destroyed all of humanity except for Noah and his descendents, cleansed the world of sin. The association of water with the erasure of sin is elaborated in the practice of baptism. Ablution and washing are an essential part of Judaism, Christianity and Islam. In all cases ritual washing requires that the water be pure. In Judaism 'living waters' prepare the individual for change and repentance. Ideally, the ritual bath or *mikvah* should be taken in rainwater. Orthodox Jews still pour water on their hands before meals and prayers or after a near-death experience (de Châtal, op. cit.: 28). In Christianity, baptism came to signify the entry of an infant into the membership of the Church, and a means of rescuing the child from sin and damnation (Oestigaard, 2005: 75). For the adult, baptism is a new birth, a resurrection. As Aquinas said, 'Baptism opens the gate of heaven.'

Christianity preserved the older pagan cults of water in the many springs and wells sacred to certain saints; holy water, blessed by a saint or priest was used to cure the ills of humans and animals. The waters of certain springs are still believed to be a source of eternal life, and churches or chapels are often built beside them. In modern Greece, the legend of Alexander the Great and the Mermaid reflects the miraculous power of water to restore life. As he lay dying, according to the tale, Alexander asked his sister to fetch him a bottle of the water of eternal life which he had brought from the east and saved for just such a moment. Mortified, his sister realized she had emptied the water without realizing what it was. When her brother died, she threw herself into the sea, but the waves took pity on her beauty and transformed her into a mermaid. All Greek children are taught this story, and no Greek pilgrim to a holy site returns without a bottle of miraculous water.

In Islam, the emphasis on the role of water in creation is perhaps stronger than in any other religion. Allah 'made every living thing from water,' the Koran states (21: 30). The equivalent phrase for the Hebrew 'living waters' is 'flowing water' in Arabic. Water, mentioned many times in the Koran, is a gift given to man by God:

> Consider the water which you drink. Was it you that brought it down from the raincloud or from Us? If We had pleased, We could have made it bitter; why, then, do you not give thanks?' (56: 68–70).

> We send down pure water from the sky that We may thereby give life to a dead land and provide drink for what we have created – cattle and men in great numbers (25: 48–49).

Elaborate rules of purification govern the everyday life of Muslims, who must wash certain parts of the body carefully before each of the five daily prayer sessions. Despite the importance of washing, the scarcity of water in much of the Muslim world meant that water should be used sparingly, and in its absence, fine sand could even be substituted for water as a means of ablution (Oestigaard, 2005: 91–3). To waste water, according to the Prophet, is a sin. In a desert climate, the value of water and the control of its sources structures both social and religious life (93). As Oestigaard notes: 'Without incorporating the desert environment and the absence of water one may not understand the development of Islam and the particular water rituals' (95).

One important difference in the cultures of the North and South of the Mediterranean is the relationship between land and water. In the North, especially the North-West of the Mediterranean, land meant power. In the South and East, water was more important than land, and was often sold together with the land (de Châtal, 2007: 46). The earliest laws we know of had to do with the regulation and distribution of water in the 'Land between the Rivers' (Mesopotamia). The careful regulation of river water for irrigation in the area was recognized in Neolithic times. In areas where river water was not available, an elaborate system of irrigation by *qanats* or underground tunnels that tap groundwater was developed in the late Bronze Age. The earliest evidence for such a system is found in Akkadian texts from the 8th century BC. The technique spread to India, China, North Africa and Spain over a long period, and is still found in Morocco. In Spain the qanats were called *madjiras*, a word from which Madrid, the city 'built over water' took its name.

In his influential study *Oriental Despotism: a Comparative Study of Total Power*, Karl August Wittfogel argued that 'hydraulic societies' like those of Egypt, Mesopotamia, China, Mexico and Peru were despotic because hydraulic construction and maintenance required slave labour (1957). Wittfogel's theory has been discredited by later scholars, who have demonstrated that the creation and maintenance of irrigation systems was more commonly the work of individual local farmers, and the role of kings and masters was to manage what had been established in a haphazard fashion.[2]

---

2 Horden and Purcell (2000: 250–51). Far from being an indication of despotic rule, the management of irrigation, according to the authors, is generally held to promote 'a cooperative social response.' They elaborate the forms these cooperative systems for

More importantly it is now recognized that all societies are, to some extent, hydraulic, even those of the driest deserts on earth. Pietro Laureano's study of traditional water techniques (2001), demonstrates how cultures from the Stone Age to the present, and from the Sahara to Southern Australia have conserved and distributed water in ways that, in some cases, have only recently been documented. It was generally assumed that hunter-gatherers simply moved on when sources of water dried up, but in Southwestern Australia, for example, it has been established that not only did the native population dig a network of ditches and embankments around their settlements using only digging-sticks, but that when this store of water was exhausted, larger social units were formed to manage the nearby marshlands by building embankments and a kind of water-labyrinth (2001: 43). Some ancient societies developed an expertise in water works that they exported or carried with them as they expanded. The Greek archaeologist Theodore Spyropoulos (1972) believed that only Egyptians could have constructed the elaborate hydraulic engineering that drained Lake Kopais allowing the ancient region of Boiotia, around Orcomenos to be supplied with water for irrigation during the summer. Taking this as evidence for an Egyptian colonization of Greece, Martin Bernal (1991) notes that the ancient Greeks themselves attributed the water system to the Egyptian King Danaus, whose name may be derived from an Egyptian word meaning water.

Whether or not Bernal is right in thinking that Egyptians colonized Boiotia in ancient times is not what concerns us here. What is striking is the central importance water had for the ancient peoples of the Mediterranean who have so dominated the history and imagination of the western world. Before looking at the modern Mediterranean and the crisis of water that is threatening the region surrounding it, it is important to consider how we might frame a discussion of the Mediterranean in a way that would justify treating it as a cultural or geographical entity.

What is the Mediterranean?

Any attempt to consider the Mediterranean as some sort of unity must begin in the shadow of Fernand Braudel's labor of unabashed love for the region. His *The Mediterranean and the Mediterranean World in the Age of Philip II* (1972) treated history and geography as inextricably linked, and the relationship of man to his environment – what he called the *longue durée* (1972: 20) – as the defining constant of human history. Braudel believed that periods of change and singular events in history took place against a slowly-moving background of human activity that was essentially unchanging and composed of recurrent patterns imposed by a particular landscape.

---

managing water took in various Mediterranean locations, but caution that there is no overall picture that fits the management of irrigated land in the region.

Since its publication, Braudel's *Mediterranean* has been the subject of many critical studies which need not be enumerated here.[3] In their often brilliant study of Mediterranean history and ecology *The Corrupting Sea*, Peregrine Horden and Nicholas Purcell note that the most obvious change that has taken place in the field of Mediterranean scholarship may be the avoidance of the entire region as a field of study (2000: 39). Their explanation for this phenomenon is partly that it is the result of actual changes in the region, and partly that it is due to a different approach to the writing of history. Interestingly, Horden and Purcell also relate the change to literary factors, noting the scholarly tradition of the 'founding fathers' of regional geography and the 'Romantic current in Mediterranean writing' (2000: 40). The very magnitude of Braudel's achievement no doubt discouraged competition, and recent approaches to Mediterranean history and geography have focused not on material life but on the psychology, beliefs and particular environments of individuals and societies.

Horden and Purcell are at pains to avoid the pitfalls of trying to address the region in conventional terms as a unity achieved by historical trade and communication (an approach they call *interactionist*) or by its physical features (*ecologizing*) (2000: 10). What defines these two approaches, they note, is an emphasis either on the sea itself or on the hinterland. Despite their inability, until relatively recently, to see the sea in its entirety, the inhabitants of the Mediterranean were aware of its existence as a large body of water as early as the first millennium BC. Navigating its often dangerous water was, they realized very early, by far the most efficient means of transporting goods from one place to another, even if ships hugged the coastline, envisaging it as a continuous stream or *periplous* (Horden and Purcell, 2000: 11). But the forces of trade and navigation did not necessarily unite the civilizations that shared its water. Rather, the sea was claimed by various peoples, especially the ancient Greeks and Romans, as their own preserve (*mare nostrum*).

It is to the credit of Horden and Purcell that despite their sensitivity to the difficulties of considering the Mediterranean as a geographical or historical unity, they have produced their own impressive Mediterranean history. What led them to undertake the project was not the broad dimension of geographical similarities or the long dimension of history, but a 'historical ecology' that takes into account the small and particular geographies, the enormous variety of micro-environments in the region and yet insists on some common threads. Interestingly, they end their volume (a second volume is projected) by affirming threads not drawn from their own respective disciplines but from cultural anthropology. Referring the culture of 'honor and shame' first adumbrated by Campbell in his 1964 monograph on a Greek Sarakatsani village, the authors conclude that despite the objections of scholars such as Michael Herzfeld,[4] these cultural values have been

3 For an extensive bibliography on the changing reception of Braudel's work, see Horden and Purcell, 2000: 541–3.

4 Herzfeld (1980, 1987: 11) argues persuasively that the emphasis on these values as emblematic of the Mediterranean is not only anachronistic, but that it places Mediterranean

maintained over a broad enough area and for a long enough *durée* to be a defining characteristic of Mediterranean society (2000: 523). A more persuasive unifying thread for considering the Mediterranean as a behavioral whole might have been hospitality. Certainly the centrality of the guest-host relationship is crucial to an understanding of much of Mediterranean literature, beginning with the Homeric poems, particularly the *Odyssey*.[5] Horden and Purcell are not so far off the mark when they say that it is possible to claim that 'the *Odyssey* has been the creator of the Mediterranean' (2000: 43). The journey of Odysseus from Troy back to Ithaca is one that links the disparate parts of the Mediterranean into a whole, an entity that can only be seen by an adventurous sailor or by the poetic imagination. On his voyage, Odysseus is dependent for his survival on the relationships he and his crew establish with their hosts. Hospitality, I will argue, is so fundamental to survival in the Mediterranean that it becomes a much more important cultural unifier than honor and shame. Indeed the links between hospitality, survival, and water underlie many of the founding myths of Mediterranean society.

The Mediterranean world has always been a world of fugitives, wanderers, pilgrims, and emigrants. On their wanderings, Mediterranean travelers from Odysseus and Jason to the Roma of Spain and the Sephardim of Salonika, have been dependent on the hospitality of strangers. An early story of enforced wandering is the myth of Europa, a tale that links the three great powers of the early Mediterranean in a family relationship. Daughter of Agenor, a Phoenician King whose father was Egyptian, Europa had the misfortune to catch the eye of Zeus. Europa transformed herself into a cow to escape Zeus's lust and fled to Crete. Zeus then changed himself into a bull to mate with her. Not only do three great civilizations of the Bronze-Age Mediterranean meet in the Europa myth; they spread out along the Nile and across North Africa to Libya. The Egyptian father of Agenor was said to be the son of the sea god Poseidon himself, and Agenor's mother Libya was the daughter of Epaphus and Memphis. Epaphus was said to be the father of both the Libyans and the Ethiopians. The myth reminds us of the dominant position Greece occupied and continues to occupy in our western imagining of the Mediterranean. Europa is a story about us; her father's foreign origins are subsumed by the success of the Olympian god in cunningly deceiving and impregnating the 'oriental' princess. To escape a lecherous god, this well-connected but foreign Europa must cross the Mediterranean and to achieve his desire the Greek god must also travel across the sea, and not only once. As he symbolically sews its shores together, he also sows his seed, uniting the Olympian deities with the great sea powers of the Mediterranean world.

---

society in a category of primitive 'otherness.'

5 There is a large bibliography on the theme of the guest/host relationship in Homer, particularly in the Odyssey. The classic text on the subject remains Stuart Douglas's 1976 book *The Disguised Guest*.

Culture Dead and Alive

Considering the Mediterranean as a cultural unity, even an imaginary one, has its advantages. It allows us to listen to what Mediterranean people say about themselves. Objecting to the sterility of old debates about what constitutes 'Mediterraneanity,' Herzfeld (2001) suggested we 'shift our attention from essentialist questions about 'what the Mediterranean is,' and move our discussion of the geographical focus from increasingly arid debates about its ontological status to a more process-oriented understanding of the politics of cultural identity, in a region where indeed some people, some of the time, find it useful to emphasize their identity as 'Mediterranean' (2001: 663). It is not only people inside the region who have found it useful to refer to the Mediterranean as a whole; it has always been viewed as an attractive alternative to the cold, industrious North. For northerners, it is a place of refuge from quotidian responsibilities, a place in which they spend their leisure time. Nostalgia for holidays spent on a sunny island or in a wine-rich village of Tuscany contributes to a rather warm and fuzzy concept of Mediterranean culture that passes for coherence and becomes a marketable commodity. The relatively recent concept of 'Mediterranean music,' for example, has been described by Bernard Lortat-Jacob as a concept for advertising or travel agents, one that offers only 'an illusion of coherence.' (2001: 539).[6]

Sociologist Gerhard Steingress uses the term *mediterraneanity* to refer to 'the living culture of the Mediterranean' (2005). 'Living culture' presumes there is a dead one and in the case of Greece, one would have to say that the value commonly assigned to its ancient cultural artifacts is considerably higher than that accorded its living ones. This emphasis on culture of the past has had unfortunate consequences, especially in Mediterranean countries like Greece, Egypt, or Italy, where the contemporary culture is often compared unfavorably to the past or largely ignored. A common view of culture as a series of 'Classical Moments'[7] has been responsible for the undervaluing of many subsequent cultures, especially in the Mediterranean area. In an attempt to redress the balance, UNESCO proclaimed the first 'masterpiece of the oral and intangible cultural heritage of Mankind' on 18 May 2001.

---

6 On the other hand, in a study with Gilles Léothaud, Lortat-Jacob makes a case for the singular, unified phenomenon of the Mediterranean voice (2002). Nasal, strongly projected, grainy (*granuleuse*), melismatic (ornamented), and with a narrow upper range, this vocal style is recognized as having acquired some of its characteristics through the process of musical acculturation. Only certain kinds of enrichment, apparently approved of by the authors, will result in true Mediterranean vocal music. More recent modifications acquired via the international media result, in their opinion, in a type of acculturation that 'ultimately risks the death of the most beautiful musics of the Mediterranean oral tradition.' (2002: 14).

7 For a discussion of this phenomenon in a comparative framework including Greece, see Holst-Warhaft, Gail and David McCann (1999).

The General Director of the UNESCO, Koïchiro Matsuura, noted that the special emphasis given to the oral and intangible cultural patrimony of humanity was the consequence of the recognition of the 'threat traditional cultures are exposed to' and, therefore, the necessity to preserve cultural diversity in the world.[8] The particular aim of the efforts to be made on this account consisted, he concluded, in the correction of the present asymmetrical situation which favors the countries of the 'North' and the protection of their material patrimony[9] at the expense of the countries of the 'South' which suffer the exclusion of their non-material cultural heritage although it is basic for their cultural diversity.

In trying to articulate a concept of Mediterranean culture that would include both its material cultural patrimony and its intangible manifestations, the work of one of the most influential theorists of culture, Pierre Bourdieu, is a useful starting point. Bourdieu identified three types of 'Cultural Capital' all of which, he claimed, confer an economic as well as a status benefit on those who possess or acquire them (1986: 47). The first type of cultural capital is *embodied* in the individual. Acquired by birth, developed through family and societal connections, and expressed through language, this type of culture situates a person in a particular *habitas*. The second type of capital exists in an *objectified* state: works of art, musical or scientific instruments are examples of objectified culture. Needless to say, their value is dependent on having acquired the first type of capital. Thirdly, there is *institutionalized* cultural capital. This is acquired through education and training and confers an economic benefit on the individual according to a society's recognition of the value of such achievements.

Bourdieu's theories have been refined in his own later works and those of his many critics and supporters. The concept of *habitas* is crucial to Bourdieu's thinking, and to understanding his definition of 'embodied' cultural capital. As King (2005: 222) points out, a person's *habitas* is not only formed by the family he or she is born into but also by the social class to which that family belongs. The accidents of class are closely related to the acquisition of the other forms of cultural capital. Bourdieu precluded the notion that he was only dealing with upper class culture by recognizing the fact that a person's cultural resources may be an advantage in one 'field' or site of social competition, and a disadvantage in another. On the other hand, as Hage's study of Australian society (1998) reminds us, in a country where a particular culture (in this case white and Anglo) dominates, individuals belonging to groups with lower status (native people, immigrants) are encouraged to exchange the cultural capital of their field, which is inevitably a marginal one, for the dominant variety in order to gain a higher position in the society as a whole.

The UN's designation of certain manifestations of 'living culture' as masterpieces seems designed to address by decree the question of the hierarchy

8 Première proclamation, p. 2.

9 Between 1972 and 2002, 690 objects were included in the UNESCO list of material patrimony.

of cultural capital. There is an implication that objectified capital already has the recognition it deserves. Like objectified capital, embodied capital is stratified but certain fields of it are underappreciated. By reifying the culture of the 'South' the UN hoped to boost its capital value. It will be interesting to see if this is more than a futile gesture in the direction of establishing a cultural parity among the world's nations. Among Mediterranean countries, the declaration may represent not only a reification of southern versus northern culture, but a recognition of the division between highbrow and lowbrow culture within a particular society.

Greece provides a good model for this distinction. Many observers have noted that Greeks have a schizophrenic attitude to their own culture.[10] No country in Europe has a more prestigious cultural heritage than Greece, and this heritage has an enormous symbolic and economic value to the modern nation. We could say that the value of its classical heritage is incalculable in terms of symbolic capital, but quite calculable in practical terms. Mass tourism, now the major industry of Greece, is the most obvious economic advantage of Greece's classical heritage, but there are others. From Greece's struggle for independence against the Ottoman Turks to the Marshall Plan and Greece's bid for membership in the European Union, Western Europe and the United States have invested in Greece as an indubitable and necessary part of the west. Not always to its own advantage perhaps, Greece has been seen as a prize to be grasped for the western block of nations at any cost. Churchill, who claimed Greece for the west at Yalta, had, after all, been educated at Eton where English upper class schoolboys were taught that a knowledge of Greek and Latin was a prerequisite for ruling the world.

On the one hand modern Greeks have been anxious to identify with what they see as their cultural patrimony: the art, philosophy, literature, and language of classical antiquity. If there is one issue guaranteed to inflame the passion of Greeks, it is the return of the Parthenon Marbles, the so-called Elgin Marbles, named after the Lord Elgin, who shamelessly stole them in the 19th century and carried them off to England. About Lord Elgin, I agree with Lord Byron, who was horrified at the rape of the Acropolis, and excoriated Elgin in his 'Childe Harold's Pilgrimage.'[11] The return of the Elgin Marbles would, Greeks believe,

10 The distinction between European and classically-derived values (the 'Hellenic' side of modern Greece), and the 'Romiot' side was first spelled out in a list of paired oppositions by Patrick Leigh Fermor (1966) and has been elaborated by many other observers.

11 Cold is the heart, fair Greece! that looks on thee,
Nor feels as lovers o'er the dust they lov'd;
Dull is the eye that will not weep to see
Thy walls defac'd, thy mouldering shrines remov'd
By British hands, which it had best behov'd
To guard those relics ne'er to be restor'd.
Curst be the hour when from their isle they rov'd,
And once again thy hopeless bosom gor'd,
And snatch'd thy shrinking Gods to northern climes abhorr'd.

'Childe Harold's Pilgrimage'.

enrich Greece's objectified cultural capital, something which is, to a large extent, measurable. A large new museum has been built to house the Marbles and other statues found on the Acropolis. It will doubtless attract millions of tourists, foreign and Greek, who will pay for the privilege of seeing the statues. More importantly, if the British Government decides to return the stolen monuments, it will restore a sense of injured pride to the Greeks and so boost the country's symbolic capital.

On the other hand, Greeks live in a country that was, until the mid-19th century, a relatively unimportant part of the Ottoman Empire, and their everyday culture is deeply affected by the centuries of Ottoman, Venetian and other domination. They may declare themselves European, but what makes Greeks feel Greek, rather than Italian or German, for example, is the culture of their everyday lives, the rituals that make up the totality of Orthodox Christianity, folk tradition and popular local culture. There is no doubt that familiarity with these rituals and skill in performing them is viewed as a form of cultural capital. If you want to be a successful politician in Greece, for example, you should be able to dance the solo male dance called *zeibekiko*.[12] This qualification for statesmanship was a product of the 1980s and 90s when the Socialist Party Prime Minister Andreas Papandreou dominated Greek politics. The *zeibekiko* may have come to Greece from Asia Minor and been associated with Turkish mercenaries known as *zeybeks*, yet this solo male dance evokes deep feelings Greeks have about themselves as a people, and the instrument that accompanies the dance, the bouzouki, is as much a metonym for popular Greek culture as the flamenco guitar is for Spanish.[13]

If one way to assess a country's image of its own culture is to look at the souvenirs that are sold in tourist shops, then Greece's cultural schizophrenia is perfectly represented by the mugs and t-shirts one can find in every souvenir shop in Athens or on a Greek island. Some have the acropolis or a classical statue on them, others a bouzouki or a Zorba-like figure dancing. Call it Apollonian or Dionysian, Hellenic or Romiot, it represents Greece culture for sale, dead or alive. Pictures of island beaches and the whitewashed villages of the Aegean are also on sale: a holiday landscape of endless summer where fresh water is nowhere to be seen.

Cultural Capital and Natural Capital

Since the mid-1970s the term 'natural capital' has been invoked by many scholars in the fields of ecology and ecological economics to refer to the natural resources available to humans as the pre-conditions for the production of man-made capital.

12 On the political and cultural significance of this dance see Cowan (1990) and Holst-Warhaft (2004).

13 Since the 1950s, Greeks have spoken of going 'to the bouzoukis' as a way to describe the establishments where the most common type of urban popular music is performed.

According to Fikret Berkes and Carl Folke, these consist of non-renewable resources such as oil and minerals, renewable resources like water and wood, and 'environmental services' such as the maintenance and regulation of the atmosphere, drinking water, climate, etc. (1993). The authors consider the concepts of natural capital and human-made capital inadequate to describe the system of interrelations between humans and the environment and introduce as an ostensibly original idea, the term 'cultural capital' (1993: 1).[14]

The authors define cultural capital, however, in very different terms from Bourdieu. In their model, it is not something that defines one's place in a hierarchical society; rather, it is a tool by which a society deals with the environment and modifies it. Cultural capital includes the way people view the world, environmental philosophy, religion and ethics, traditional knowledge, traditional ecological knowledge, and social and political institutions. Recognizing the inadequacy of the term 'cultural capital' for their systems approach to sustainability, Berkes and Folke also introduce the term 'adaptive capital' to stress their interest in all factors that concern ecological economics 'from an evolutionary, mainly cultural evolutionary, sense' (2).

From a systems perspective, the environment is viewed as providing a limit on the growth of the biological subsystem. Organisms are constantly adapting to the environment and the ecological system exists in a process of self-organization, the water-cycle being a perfect example. The environment places physical constraints on human development, causing humans to adapt and modify their activities. Each society's way of organizing its resources and its relationship to the particular environment in which it finds itself is unique. Recognition of the need to study these diverse and traditional techniques of adaption in order to conserve dwindling or threatened natural resources has been relatively recent. In 1992, at the Rio de Janeiro conference on the Environment and Development, the United Nations recognized the necessity for an inventory of traditional knowledge. The UNCCD (the Convention to Combat Desertification and the Degradation of Soils) founded a Science and Technology Committee whose task was to compile a list of techniques and practices organized under seven headings, the first of which was 'water management for conservation' (UNCCD, 1998a). In their report, the scientists charged with carrying out this task defined traditional knowledge not only in technical terms but as a set of holistic practices 'handed down from generation to generation and culturally enhanced' (ibid). In his handsomely illustrated book *The Water Atlas* (2001) Pietro Laureano describes how the Italian Ministry for the Environment approached the task of producing its inventory, recognizing the complexity of traditional practices. He takes the example of terracing, one of the commonest Mediterranean practices used to protect slopes, retain water and replenish soil. Besides its functional effectiveness, Laureano insists that the

14 'Moreover it is not possible to approach sustainability by only focusing on these two factors, natural capital and human-made capital interrelation. We need a third dimension, what we refer to as cultural capital as well.'

terracing took on an aesthetic and social value (2001: 24). In contrast to modern technological methods of harvesting and managing resources, which operate 'by separating and specializing,' local knowledge does not distinguish between forest, tilled land and dwellings since all coexist in constant, productive interaction. But there is more to traditional knowledge than utility. The photographs and diagrams in Laureano's book emphasize the aesthetic, sacred, and ethical dimensions of the interaction. He notes that in the Sahara, as in southern Italy, small agricultural fields are called *gardens*, a term that encompasses fertility and pleasure, and systems of Saharan water distribution, which are part of the complex symbolism of life and fertility, and are reproduced in carpet designs and women's hairstyles (26).

'Traditional knowledge' is not a series of techniques that can be adapted to fit modern needs. Understanding the way hunter-gatherers, farmer-breeders or metal-using agro-pastoralists used and conserved their environment is obviously not always relevant to the complexities of modern urban life, but, as Laureano points out, the desert oasis, with its scarce water carefully managed, and the urban ecosystem are not so far apart. Small oases grew into great caravan cities. The defining feature of these larger agglomerations of people was a form of capital: 'isolated basins in the middle of the desert, large plains among the mountain peaks, strips of oases along hydrographic networks; international and intercontinental crossroads' (32). It is Laureano's belief that the traditional knowledge that made life possible in an unpredictable environment such as the Mediterranean can be used as a paradigm to address the present planetary emergency 'caused by the water crisis in the form of an excess or scarcity of water, soil erosion and environmental and urban collapse' (367).

A Greek Song and Hospitality

My own involvement with the problem of water in the Mediterranean was prompted by a visit to an island I had lived on for a year and by a Greek song composed in the 1960s by Mikis Theodorakis with words by the playwright Iakovos Kambanellis:

> The bread is on the table,
> The water's in the jug,
> The jug is on the staircase,
> Give the thief a drink.
>
> The bread is on the table,
> The water's in the jug,
> The jug is on the staircase,
> Give Christ a drink.

Mother, give to the passer-by,
To Christ and to the thief,
Give him his fill, Mother,
Give him a drink, my love.

The words reminded me of the Greece I had discovered many years ago, a largely rural Greece, where hospitality was extended to every passer-by, regardless of their need or creed. And hospitality meant bread (or some home-baked equivalent) and water, cold water drawn from the well. Almost all rural Greek households in the 1970s had a well. Many also had rain-water cisterns that were used for animals or for watering the kitchen garden. The quality of water was discussed and compared from village to village in much the same way as olive oil was judged. Listening to the song thirty years after I first heard it, I realized that it sounded like a piece of ancient history. On the island of Aegina, where I had lived for more than a year, there was no longer any fresh water fit to drink, even to water kitchen gardens. The ground water was infiltrated with salt. A tanker brought water to the island each day in the summer, water that was used for most household purposes. Islanders and visitors drank bottled water, the beaches were littered with plastic bottles that were not recycled, and nobody I spoke to seemed to think there was a water crisis. The main crop of the island, pistachio nuts, had fetched high prices when I lived on the island, but demanded more water than the aquifer could support. Ancient olive trees and almonds had been felled en masse in the 1950s and 60s in favour of the non-native, upstart pistachios, partly because the profit on the crop was higher, and because pistachio nuts could tolerate a relatively salty soil. They also needed a lot of water that could still be found in the 1970s provided you dug deep enough.

Because of its proximity to Athens, Aegina not only has the usual summer invasion of foreign tourists, but a large population of weekend visitors from Athens, many of whom have built second homes there. Despite the pollution of the aquifer and an increasing demand for water, housewives continue to hose down the floors and the pavement in front of their houses in summer, cars are washed in the noonday sun, crops are watered during daylight hours, and there is almost no recycling of water. The reason for the profligate use of water in Greece is both cultural and political. Water is supplied free to farmers in most parts of Greece and householders are charged a minimal amount for water. Despite increasingly high temperatures,[15] spray irrigation is still practiced in many parts of Greece, and there is little incentive to mend leaking irrigation pipes. It would be a bold politician who would advocate imposing a charge on water used for irrigation. The rural vote has been courted with substantial subsidies by the major political parties since democracy was restored to Greece in 1974, and farmers regard water as a right. To find it, they have drilled deeper and deeper to find water on their land, and added fertilizers to increase the yield. The result has been wholesale pollution of the

15 According to environmentalists, summer temperatures in Greece are expected to average 41°C by the year 2070 (Digital Journal, 2007).

aquifers and rivers of Greece. What agriculture began, industry and urbanization has completed. Despite the plethora of regulations introduced by the EU to prevent such destruction, Greece has been notoriously non-compliant with them.[16] A combination of agricultural and industrial pollutants has destroyed many of the rivers and wetlands of Greece, and still Greeks are reluctant to admit there is a water crisis, either in their aquifers or in the blue salt water of the Mediterranean, now a severely compromised sea.[17]

Greece is by no means unique in its misuse of water. Despite the millennia of respect for and conservation of water in the area, almost all the Islamic countries of the Middle East and North Africa face a crisis of water quality and quantity. Climate change may be one factor affecting the crisis, but there are social and cultural factors that are more important. One is the enormous increase in population, much of it urban.[18] Another is that the very traditions that placed high value on water have encouraged what Daniel Hillel calls a 'culture of waste' (1994: 214). As in Greece, water has traditionally been supplied free or at minimal cost to farmers in the region. This has encouraged the cultivation of water-intensive crops like sugar-cane and rice in Egypt, or bananas in Jordan, Lebanon and Israel (Hillel, 1994: 213–14). Traditional agricultural practices in these countries seem to be standing in the way of water conservation; on the other hand, the laws of Islam are being used by some Muslim countries to promote the conservation of water. Mohammed and his followers enjoined thriftiness in the use of water, and Islamic law (enshrined in the *Shariah*) is explicit on the need to share water equitably. The two principles of the 'Shi'at al-ma,' the laws governing water, ensure the right of thirst for people and animals and the right to water one's crops. Now the Egyptian, Jordanian and Syrian governments quote the Koran in their efforts to promote water conservation (De Châtal: 170).

Fresh water is a casualty of the phenomenon that took place in many countries around the Mediterranean in the second half of the Mediterranean. In a single generation, countries moved from being largely rural to predominantly urban. The traditional customs of hospitality that were linked to a particular landscape and

16 There have been many articles in the Greek Press in recent years documenting the continued pollution of Greek aquifers, including the illegal drilling for water in the Argive Plain and the dumping of toxic waste into the Asopos River. In the year 2007, for example, all national newspapers, especially *To Vima*, *Eleftherotypia* and *Ta Nea* ran articles on the situation, many of them critical of the government for its failure to prosecute illegal dumping and drilling. An article published in *To Vima* in 2003 noted that the European Committee had criticized the Greek authorities for failing to control levels of nitrate in various regions of the country (2 March 2003, p. A 45).

17 For a recent report from the World Bank on the state of the Mediterranean Sea, see the Bank's Annual Review (July 2007–June 2008), p. 33.

18 According to the World Bank, the population of the Mediterranean's coastal states has doubled in the last 40 years and is expected to reach 600 million by 2050. Exacerbating the increase is the summer invasion of tourists; by 2025, the number of tourists visiting the region is expected to reach 350 million.

history changed. In most northern Mediterranean cities, people turn on a tap and out of it, for at least some hours of the day, flows clean water. Women no longer fill buckets of water at the well, nor do they carry heavy pots of it on their heads back to their houses. When they did, they had a very different relationship to water. They guarded it carefully and recycled the water they washed their clothes in to water their gardens. It was a relationship of respect, of conservation, but it went further than that. It is no accident that one of the strangers who pass by the house, in that Greek song, should be Christ. As we have seen, water was a sacred substance in all three of the contemporary religions that originated in the arid lands of the eastern Mediterranean. Islam, Christianity and Judaism absorbed and codified ancient beliefs about and attitudes to water, beliefs that grew out of a particular landscape where water was a scarce resource.

I was touched, on a visit to Cairo in 2007, to see a jug of water hanging outside a small basement workshop in a narrow street in the city with a message scrawled above it in Arabic telling passing strangers to help themselves. 'Give the thief a drink, give Christ a drink'.... Hospitality asks no questions of the stranger, and in the lands of the eastern and southern Mediterranean, hospitality has always been related to water. The jug on the wall was a touching reminder of tradition, but it was a small gesture in a transformed world. The World Bank estimates that twenty million people who live along the southern and eastern shores of the Mediterranean have no access to safe drinking water, and 47 million have no access to sanitation. By the middle of our century, these figures are likely to quadruple.[19] What can culture do in the face of such a threat? Is there any way to invoke tradition in order to conserve and protect what is left of the Mediterranean's 'natural capital'?

The Imagined Mediterranean

The Mediterranean as a coherent region may only exist in the imagination of writers, painters, musicians, philosophers. Why this should be so is a long story, but since our interest is in water and how the present crisis in Mediterranean counties might be ameliorated, it is worth considering how an 'imagined Mediterranean' might come to the aid of the real thing. After all, the Mediterranean is the only region of the world, to my knowledge, that has inspired people from a dozen countries to volunteer in the armed struggles of two of its nations for freedom: 19th century Greece and 20th century Spain. In both cases the motivation for young men from as far away as England and the United States to serve in these wars was largely imaginary.

Lord Byron, the most famous volunteer in the Greek War of Independence, had been, like most foreigners who joined the Greek cause, educated in what continue to be called 'the Classics.' Inspired by the literature they read and convinced that ancient Greece was the source of what was most precious in their own culture,

---

19 'Toward Sustainable Development,' 2008 annual review.

19th century Philhellenes saw the Greek landscape as a sacred *topos* of their own imagination:

> Where'er we tread 'tis haunted, holy ground;
> No earth of thine is lost in vulgar mould.
> But one vast realm of wonder spreads around,
> And all the muses' tales seem truly told,
> Till the senses ache with gazing to behold
> The scenes our earliest dreams have dwelt upon.[20]

Philhellenes from all over Europe and from the United states were prepared to die, far from their native countries, for the liberation of a land they knew nothing about except what they had learned in their schooldays. Most of them had little knowledge of the contemporary Mediterranean, and most had acquired their vision of Greece from classical literature. This imagined landscape, and the vision of it created by contemporary English poets like Byron, and Shelley, and by their American equivalents Fitz-Greene Halleck, James Gates Perceval, and William Cullen Bryant, inspired thousands of young men to fight in the Greek cause and may have been responsible for the ultimate Greek victory over the Turks.[21]

A century later, inspired by a different but equally unrealistic vision of the contemporary Mediterranean, thousands of northern Europeans and Americans flocked to Spain to join the anti-fascist forces. Although this was an ideological struggle rather than a battle for independence, something more than ideology attracted many of the volunteers to fight and die in a foreign country. Even George Orwell, whose disillusionment with the Spanish Republican and Communist leadership was rapid and absolute, described his attraction to Spain in an uncharacteristically romantic passage:

> I seemed to catch a momentary glimpse, a sort of far-off rumour of the Spain that dwells in everyone's imagination. White sierras, goatherds, dungeons of the Inquisition, Moorish palaces, black winding trains of mules, grey olive trees and groves of lemons, girls in black mantillas, the wines of Malaga and Alicante, cathedrals, cardinals, bull-fights, gypsies, serenades – in short, Spain. Of all Europe it was the country that had most hold on my imagination. (1962: 194).

This hold on the imagination is not an ephemeral one. For better or for worse it is remarkably persistent and partly explains the consistent and ever-increasing volume of tourism in the Mediterranean region. In her perceptive study of the interrelation of culture and geography, Artemis Leontis suggests that the relationship between the geography of Greece and its mythological or literary associations can

20 Byron, 'Childe Harold's Pilgrimage'.

21 On the American Romantic poets and their influence, see Holst-Warhaft, 2007: 143–156.

have important physical consequences (1995: 20–21). Leontis takes the fate of the Illisus River as an example. Quoting a study by Panayotis Turnikiotis, she notes that the river became a site where modern Athenians strove to recover a legendary past. Plato's allusion to the river as the mythical stream where 'Boreas seized Oreithya' (229b), and its banks as a place that inspired Socrates' famous disquisition on love, made the river a quasi-sacred *topos*. Pausanias' guide to the ruins of ancient Greece enhanced the reputation of the river, making it a site no traveler could afford to ignore.

Whether on not the river ever displayed more than a trickle of water, by the time the Greek capital was relocated to Athens in the 1850s, it had run dry. Its polluted bed divided the more prosperous from the poorer sections of the small city. Because of its illustrious associations, though, efforts were made for almost a century to locate the ancient stream-bed, clean up the pollution and establish some sort of cultural activity along its banks. These efforts were ultimately defeated by the needs of urban development. In the 1950s city officials agreed to fill the gully in order to create the new King Constantine Avenue. The tale of the Illisus suggests that however unlikely the recovery of the ancient river was as a modern refuge from the city, Greek town-planners were willing to make a considerable effort over the course of a hundred years to restore the physical site of the river's dry bed. The power of the intangible and tangible inheritance, of history and myth transmitted through literature, song, and other works of the imagination, to effect change on the physical landscape, particularly a river, may be a fruitful model for environmental change not only in Greece but in the broader Mediterranean region. The region's cultural heritage can and has been harnessed for political or nationalist purposes. It can also be appropriated to help solve the present crisis of fresh water.

As Gaspar Mairal makes clear in his contribution to this volume (ch.2), a national water policy may have as much to do with imagination as it does with practical considerations. In the case of Spain, a grandiose plan for bringing water to the arid parts of the country was conceived at a time of national defeat and seemed to offer a way out of the crisis, but as in the US, and the USSR, Spain's ambitious hydraulic policy had mythical dimensions:

> The USA, the USSR and Spain all provide examples of how ideology, technology and power combined in a water policy based on the construction of huge dams involving immense scientific and technical challenges, and the investment of untold sums of money. Yet all of this effort was inspired, driven and directed by myths, some Biblical in origin, such as the need to colonize a 'promised land' and to create a garden in the midst of the desert, and others Communist, such as the construction of the 'new man' by the socialist state (Mairal, 2009).

Mairal demonstrates that Spain's policy was initially conceived not by engineers but by a late 19th century jurist and politician named Joaquín Costa whose goal was not simply to reform his country's water policy but to revive his demoralized nation. The success of his grand vision was dependent on his ability to spread his message of redemption by water to Spanish farmers and to the public at large. His language, rich with Biblical metaphors, envisaged a Europeanized Spain, transformed from a desert into a flowering garden. The myth of the transformation of the desert into a garden is a common one in the Mediterranean, and central to its three major religions. Costa's ability to influence public opinion in favour of an ambitious water policy was precisely the familiarity of his audience with such myths.

In the contemporary Mediterranean, Christian, Jewish and Islamic metaphors have all been invoked in the context of nation-building and public policy. In the face of the crisis of fresh water which confronts most of the countries of the Basin, can such mythical rhetoric be effective? I would argue that political rhetoric has been discredited in many Mediterranean countries. Too much has been promised, too little delivered. On the other hand culture, in its traditional and local manifestations, has been a constant source of national strength that has both transcended politics and been enlisted to support it. This may be why the leading country in water policy in the Mediterranean has been so successful. Spain's 'New Water Culture,' grew out of a combination of academic discussion and social movements that began with protests against large dam schemes in the mid-1990s. Controversy over water transfers from the Ebro River led the Ministry for the Environment to publish a 'White Book' on water in Spain in 1997.[22] The following year a 'New Water Culture Foundation' was formed with a membership of one hundred, most of them academics. Their focus was on conservation and technical debate, but they also saw themselves as a motivating force for social change.

The level of debate, public involvement and policy development achieved in Spain is unique, and its success may be attributed, in part, to its emphasis on culture. In its manifesto, the New Water Culture movement emphasizes the fact that the use of the word culture assumes a change of paradigm that views water not as a product but as:

> an eco-social asset in which the root 'eco' recuperates the broader Aristotelian term 'oikonomia' – the art of managing the household, with the double aspect of economic and ecological management (Mairal, 2009).

Whether the Spanish Water Culture should be considered a cultural phenomenon or a political movement may be debated, but in their rhetorical emphasis on 'culture' rather than 'policy,' its leaders have been successful in distancing themselves from conventional political disputes over water. Given the importance of culture as a motivating force in the region, they have provided a model for what might be successful in other Mediterranean countries: the recuperation of the relationship between water and culture as a means to address the escalating water crisis in the

22 Mairal, lecture on the 'New Water Culture' of Spain, 9 March 2009, Cornell University.

region. If the word culture is more effective than policy or politics, why is the rest of the Mediterranean not using it to resist the degradation of its water resources? The answer is probably specific to each single country, and to its political and economic condition. Invoking the 'culture of water' in a country with a galloping urban population and little sanitation seems a luxury, but it may be one that we cannot afford to do without. When governments are unable to satisfy the needs of a population for water, the result is either political collapse or forced migration. The economic costs of both are much higher than any costs associated with conserving, protecting, and distributing scant water resources. All this is known, but it does not prevent the pollution of rivers or stop the illegal drilling of deep wells in agricultural areas. Using culture as one weapon in the struggle to improve the quality and quantity of available water in the Mediterranean may not be a dream, but a vital task. In central Australia 'water dreaming' has always been considered a vital principle of the life cycle for native people. The dream of an imaginary Mediterranean world encouraged armies of foreigners to fight in real wars. Water dreaming may be as effective a means to change the fate of the Mediterranean as any other. The case of the Asopos River in Greece would seem to be a perfect opportunity for using culture, rather than policy, to rescue a dead and deadly body of water.

Tale of a River: Dreaming in Hell?

The Asopos River, which forms a natural border between Attica and Boiotia is perhaps the most egregious example of fresh water pollution in Greece. In 1969 the river was officially designated as a dumping site for the discharge of industrial waste from the many factories in the area. Local residents began protesting about the putrid, possibly toxic waters of the river in 1974, but despite their concerns, and the declaration of the river as a protected wetland by the European Community, the entire length of the river was pronounced a dumping site by the four prefectures that bordered it in 1979. It was not until 1994 that the river water was tested for chemical pollutants by researchers at the University of Athens. It took them another two years to conclude that the level of industrial dumping was excessive. There was still no mention of the danger the water might pose to the inhabitants of the nearby town of Oinofyta. In 1997 the Ministry of Health set up a committee for the protection of the Asopos which took no action. Meantime residents concerned about what seemed to be escalating incidences of cancer in the neighborhood sued thirteen industries for pollution, but the fines imposed in the settlement were extremely low. The Association of Greek Chemists agreed that chromium 6, lead, nitrates and other pollutants have reached alarming proportions in the Asopos, yet the Greek Government still denies that the river poses a health risk. The residents now drink bottled water, but the river is used to irrigate crops, most of them root vegetables intended for the markets of Athens. On her website, the American environmental activist Erin Brockovich noted that the levels of chromium 6 in the

Asopos were much higher than those she famously reported in the Hinkley River in California (2007).

In the summer of 2008, a small group of researchers from the US and Mediterranean countries visited Oinofyta with Panayiotis Nektarios, a researcher from the Agricultural University of Athens. We were welcomed by the Mayor and the local priest, Father Yiannis, a longtime activist against industrial dumping. When we drove to the Asopos to look at the sluggish trickle of blood-red liquid flowing between its banks, we were followed by a local farmer in an SUV. He urged us not to believe what they said about the river. There was nothing wrong with it, he assured us. The recently-elected mayor had intended to have a public meeting to discuss the situation with us, but fearing unwelcome interruptions, he closed the door of the town hall and told us his dream.

> When I was a child, my mother would bring me down to the river to pick flowers, and the water was clear. My dream is to be able to bring my children to the river to pick flowers again.

The Asopos is a river as rich in ancient mythical associations as any in Greece. It is associated with the Boiotian cult of muses and nymphs. Asopos, an important river god, was a son of Okeanos and husband of the water nymph Metope. If the Illisus River could inspire the founders of modern Athens to invest 100 years of effort rescuing its dry bed, the still-flowing Asopos deserves recognition as a mythic as well as a modern site. It has all the cultural ingredients needed to inspire a national, even an international dream.

Conclusion

The right to water may one day be formulated as a piece of international law but that will not guarantee the thirsty get water to drink or that the water they drink will be clean. The countries that border the Mediterranean are facing a water crisis that few of them seem willing to confront. Attempts to control population, development, or the dumping of raw sewage and industrial waste into the sea and rivers of the region have been ineffective and uncoordinated.[23] Considering water as a form of 'natural capital' is one way to emphasize its value, and in some countries it may be an effective strategy but in the face of government and societal neglect or unwillingness to act, neither international legal regulations, nor economic models of 'capital' are enough to change people's attitude to water. Something more imaginative is called for. It is clear that water has a deep cultural and religious importance for all the countries that border the Mediterranean. For historical, cultural and geographical reasons the region also has a unique ability to inspire not only its own populations but its myriad admirers to act on its

23 World Bank report, 2007–8.

behalf. These precious, non-material assets can and must be part of a campaign to educate and inspire citizens, children, and the 300 million tourists who visit the Mediterranean each summer to take up the cause of fresh water as the most serious challenge the sea and its coastal nations have ever faced. Water dreaming may be one of the most powerful weapons in the struggle to save the imperilled water resources of the Mediterranean.

References

Berkes, F. and Folke, C. 1993. A systems perspective on the interrelationships between natural, human-made and cultural capital. *Ecological Economics*, 5(1) 1–8.

Bernal, M. 1987–1991. *The Afroasiatic Roots of Classical Civilization*. Vol. 2. London: Free Association Press.

Bourdieu, P. 1986. The forms of capital. in *Handbook of Theory and Research for the Sociology of Education*, edited by J. Richardson. New York: Greenwood, 241–258. Originally published as Ökonomisches Kapital, kulturelles Kapital, soziales Kapital in *Soziale Ungleichheiten (Soziale Welt, Sonderheft 2)*, edited by Reinhard Kreckel. Goettingen: Otto Schartz & Co. 1983.183–98.

Braudel, F. 1972–3. *The Mediterranean and the Mediterranean World in the Age of Philip II.* 2 vols. New York: Harper and Row.

Brockovich, E. 2007. River runs purple and contaminates Greek town. Available at www.brockovich.com accessed 10.12.2007.

Campbell, J.K. 1964. *Honor, Family, and Patronage: A Study of Institutions and Moral Values in a Greek Mountain Community.* Oxford: Oxford University Press.

Cowan, J. 1990. *Dance and the Body Politic in Northern Greece*. Princeton: Princeton University Press.

De Châtal, F. 2007. *Water Sheikhs and Dam Builders: Stories of People and Water in the Middle East.* New Brunswick and London: Transaction Publishers.

Digital Journal 2007. Environmentalists warn Greece of climate change risks. Available at http:www.digitaljournal.com/article/226196/.

Fermor, P.L. 1966. *Roumeli*. London: Faber & Faber.

Hage, G. 1998. *White Nation: Fantasies of White Supremacy in a Multicultural Society*. Annandale, Australia: Pluto.

Herzfeld, M. 1980. Honor and shame: problems in the comparative analysis of moral systems, *Man* 15, 339–51.

Herzfeld, M. 1987. *Anthropology through the Looking Glass: Critical Ethnography on the Margins of Europe.* Cambridge: Cambridge University Press.

Herzfeld, M. 2001. Ethnographic and epistemological refractions of Mediterranean identity, in *L'anthropologie de la Méditerranée*, edited by D. Albera, A. Blok and C. Bromberger. Paris: Maisonneuve et Larose. 663–683.

Hillel, D. 1994. *Rivers of Eden: The Struggle for Water and the Quest for Peace in the Middle East.* Oxford: Oxford University Press.

Holst-Warhaft, G. 2004. The body's language: representation of dance in modern Greek literature in *Greek Ritual Poetics*, D. Yatromanolakis and P. Roilos (eds) Center for Hellenic Studies.

Holst-Warhaft, G. and McCann, D. 1999. *The Classical Moment: Views from Seven Literatures.* Lanham, MD: Rowman and Littlefield.

Horden, P. and Purcell, N. 2000. *The Corrupting Sea: A Study of Mediterranean History.* Oxford: Blackwell.

King, A. 2005. Structure and agency theory, in *Modern Social Theory: An Introduction.* Austin Harrington (ed.) Oxford: Oxford University Press. 215–232.

Laureano, P. 2001. *The Water Atlas: Traditional Knowledge to Combat Desertification.* Turin: Bollati Borrighieri.

Leontis, A. 1995. *Topographies of Hellenism: Mapping the Homeland.* Ithaca: Cornell University Press.

Léothaud, G., and Lortat-Jacob, B. 2002. La voix méditerranée; une idenité problématique, in *La vocalité dans les pays d'Europe méridionale et dans le bassin méditerranéen.* La Falourdière: Modal. 9–14.

Lortat-Jacob, B. 2001. S'entendre pour chanter: Une approche anthropologique du chant en Sardegne, in *L'anthropologie de la Méditerranée*, (ed.) D. Albera, A. Blok and C. Bromberger. Paris: Maisonneuve et Larose. 539–554.

Oestigaard, T. 2005. *Water and World Religions: An Introduction.* Bergen: University of Bergen.

Orwell, G. 1962. *Homage to Catalonia.* Hammondsworth, UK: Penguin Books.

Plastino, G. 2003. Open textures: on Mediterranean music, in *Mediterranean Mosaic: Popular Music and Global Sounds,* (ed.) Plastino. New York: Routledge. 180–191.

Spyropoulos, T. 1972. Aigiptiakos epoikismos en Boiotiai. *Archeological Proceedings.* Athens: Archeological Institute.

Steingress, G. 2005. Mediterraneanity as cultural heritage: politics with the past. *Scripta Mediterranea.* XXVI.(3). 3–19.

Stewart, D. 1976. *The Disguised Guest: Rank, Role and Identity in the Odyssey.* Lewisburg, PA: Bucknell University Press.

UNESCO, 2001. *Première proclamation.* 18 May 2001.

UNCCD Convention to Combat Desertification and the Degradation of Soils Report, 1998a.

Wittvogel, K.A. 1957. *Oriental Despotism: A Comparative Study of Total Power.* New Haven: Yale University Press.

Chapter 2
Water Policy as a Gospel of Redemption

Gaspar Mairal

> Only the Bible, however, is fundamentally concerned with the macropolitics of regimes – the grand politics that guide and shape the behaviour of leaders and followers.
>
> Aaron Wildavsky, *The Nursing Father: Moses as a Political Leader*

Water Policy

Public policy has become a significant field of research in the area of Cultural Anthropology. Policy brings the wider world into contact with the local, and it is because of this that Anthropologists have stressed the local in their analyses. The ethnographic tradition of the discipline has reinforced this local view by revealing the weaknesses of policy measures that fail to take account of the cultural dimension and social impact of mega-development projects in specific communities and territories (Cernea 1991). A different approach also exists, however, which seeks to identify the cultural foundations of public policy. It is quite wrong to assume that public policy, based as it is on expert thinking, only has room for scientific contributions, relegating cultural input to a marginal status, a discretionary matter at most. According to this view, culture is considered a matter for the people and science and technology for the experts, for large organizations and for government. I take a different stance, however, seeking to show how the cultural underlay of public policies hybridizes scientific, technical and cultural concerns. Water policy is an outstanding example, and I shall base my argument on the manner in which development in this area was managed in Spain in the late 19th century and throughout the 20th, at the same time establishing a significant comparison with the situation in the United States. This comparison is based on a coincidence of historical circumstances.

At the end of the 19th century, the USA, having won the West, found itself in possession of vast, arid lands that could only be dominated by transforming millions of acres into fertile farmland, a goal that could only be achieved by irrigation. It was the Mormons of Utah who took the lead in making a garden of the desert. Most of the territory of Spain, meanwhile, belongs to the Mediterranean basin. The climate is dry and the land arid, and agriculture has always been constrained by drought and limited crop yields. The defeat suffered by Spain at the hands of the USA in the Cuban War, which concluded with the treaty of Paris in 1898, put an end to the Spanish colonial empire. This in turn triggered a national crisis as the country's elites tried to come to terms with their awareness of Spain's decadence, which by this time had reached a nadir. If the USA was now an expanding great power, Spain was a shadow of its former glory. In both cases, however, water policy emerged as an engine of change. In the United States it provided an opportunity for the nation's expansive surge as the West was colonized, while in Spain it offered the chance for an exhausted nation to regenerate and emerge from the depths of its crisis.

The expansion of irrigation as a means to transform millions of hectares of land into farmland became a key development policy all over the world in the 20th century, particularly in the developing countries. However, the first steps in this direction were taken decades earlier in the USA and Spain. The first Federal plan for the development of new irrigation schemes in the USA dates from 1902 (Reisner 1986, 2). In the same year the Spanish parliament approved its first *Plan de Construcciones Hidráulicas*[1] the objective of which was to irrigate over one million hectares (Díaz-Marta 1998). Other countries would soon follow suit, like the Soviet Union, which embarked upon vast irrigation schemes that were finally abandoned only in the post-Soviet era. Israel too made enormous progress in the extension of irrigation to the desert using new technologies. Currently, China and India are the leading powers initiating major dam and irrigation schemes in a world where such policies have increasingly been revised and even criticized (World Commission on Dams 2000).

In this chapter, I shall try to cast light on some of the myths encumbering the new water policy, or *política hidráulica*[2] as it is generally called in Spain. The contribution of expert knowledge to this policy in the form of hydrogeology, hydraulic and civil engineering, as well as other technical and scientific disciplines made a good fit with the myth, eventually forming a symbiotic relationship with it. This marriage may seem paradoxical, but it was common practice throughout the 20th century as the policy took shape. The myth sustains and orients a policy that is defined by great technical challenges and the deployment of enormous human and financial resources. How did this paradox come about? The Soviet Union provides a key example, although it is one that has been largely ignored. As Stalinism took hold, the myth of the "new man" inspired a water policy of truly gigantic dimensions[3], which would not be abandoned until first the *perestroika* and finally the collapse of the Soviet Union brought it to an end. Frank Westerman's excellent book *Engineers of the Soul* provides a fascinating account of this policy, especially with regard to *perebroska*[4], the policy of diverting rivers. As late as 2001 in an interview conducted by Westerman, Alexandr Velikanov, an engineer at the Moscow Institute for Water Policy, would complain bitterly that this "policy"

1 Hydraulic Infrastructure Construction Plan.

2 Hydraulic Policy.

3 Stalin in fact began to organize the GULAG in the early 1930s in order to build the White Sea Canal in record time. During the Franco dictatorship in Spain, political prisoners were also used to construct dams and irrigation works.

4 The policy of diverting rivers.

had been laid aside in the new post-Soviet Russia. Westerman describes his conversation with Velikanov, to whom he attributes the following remarks:

> If we can transport oil and gas over vast distances, why cannot we divert water from one river to another? The Americans do it. *Inter-basin water transfer.* It's nothing new. They have been doing it for years. Haven't you heard of the Colorado River? Or the Tagus in Spain? The only difference is that our project was on a different scale. The very size of it aroused respect. Sometimes even fear (Westerman 2005, 245).

In 1977 Brezhnev had ordered work to begin on the "Southern Strategy", a plan to divert five rivers in European Russia and Siberia designed to transfer 60,000 hm^3 of water per year. The collapse of the Soviet Union meant the end of this gigantic water policy. The USA, the USSR and Spain all provide examples of how ideology, technology and power combined in a water policy based on the construction of huge dams involving immense scientific and technical challenges, and the investment of untold sums of money. Yet all of this effort was inspired, driven and directed by myths, some Biblical in origin, such as the need to colonize a "promised land" and to create a garden in the midst of the desert, and others Communist, such as the construction of the "new man" by the socialist state[5].

Much of Spain and the western United States is characterized by the aridity of the landscape. California is perhaps the nearest comparable State with some 1,400 dams of over 7.5 meters in height and a storage capacity of 52,424 hm^3. Spain has 1,102 dams of over 15 meters with a total storage capacity of 52.710 hm^3. Meanwhile, annual water flow in California is 95,405 hm^3 (or 104,805 hm^3 including the transfer of 9,400 hm^3 per year from the Colorado River) compared to an annual 112,750 hm^3 in Spain. California has an area of 400,000 km^2 against a little over 500,000 km^2 for Spain (Arrojo and Naredo 1997). In light of these data, a meaningful comparison can be drawn between Spain and California and, indeed, between the investment made in water resources and infrastructure in both territories over the course of the 20th century. However, the discourse underpinning this enormous outlay was couched in strictly national terms and should be analyzed as such. This is what makes a comparison between Spain and the United States so interesting. While the size, population and wealth of the two nations differed markedly, the protagonists, myths and discourses of water policy are in many ways strikingly similar in this historical period.

Joaquín Costa was a late 19th century jurist and politician whose goal was to bring about reform and the regeneration of Spain. He was intensely active in

5 The myth was propagated through the new literature encouraged by the Soviet regime under the strict rules of "Socialist realism" led by Maxim Gorki. This literature would sing the praises of the tremendous engineering works undertaken in the Stalinist epoch. In 1935, for example, 36 writers including Gorki himself published *Belomor*, a work extolling the construction of the White Sea Canal by Stalin in record time.

the promotion of dams, canals and other water infrastructure projects to extend irrigation and modernize Spanish agriculture. Though Costa died in 1911, his ideas were to have a major impact on the water policy that was eventually implemented in 20th century Spain. His political discourse was shaped around an idiosyncratic agrarian populism, but he was far from harbouring any revolutionary objective and he at no time envisaged the peasantry as the protagonists of social upheaval. What we find in Costa is rather a focus on "water policy" (or *política hidráulica* as he termed it)[6], which was intended to produce reformist outcomes that would resolve the crisis of traditional agriculture at the end of the 19th century.[7] Can this "hydraulic" thinking be compared to the discourse found in any other part of the world? The question is highly relevant if we are to place Costa's work in the context of water policy in the wider world, and my research in this area has led me to consider John Wesley Powell, a significant, if colourful, figure in US history. In my opinion, there is much that is common in the history of water development in Spain and the American West, while Costa and Powell were alike in at least two ways.

Joaquín Costa was born in 1846 and died at the age of 65 in 1911. Powell was born in 1834 and died at 68 in 1902. The two were, then, contemporaries. Costa was a jurist and Powell a geologist, but they both had wider interests and were active in other fields. The anthropological dimension of Costa's work, revealed in his study of the traditional ways of the Spanish peasantry, has its counterpart in Powell's documentation and descriptions of the languages and livelihoods of the American Indians. It was Powell who founded the Bureau of Ethnology attached to the Smithsonian Institution. Both men were devoted practitioners of fieldwork. Though Costa could not be called an explorer, given that all of Spain had long since been charted by this time, he travelled widely around the country contacting local people to collect first-hand information about the traditional common law and agricultural practices of its regions. He wrote two books based on this research: *Derecho Consuetudinario en el Altoaragón*[8] (1880) and *Colectivismo Agrario en España*[9] (1898). Powell, on the other hand, was an actual explorer and the leader of one of the last great expeditions in the USA, a journey up the Colorado River, which he rafted for the first time in 1869. His account of this journey, *The Canyons of the Colorado,* was published in 1895, confirming his reputation as the discoverer of the Grand Canyon. Both Costa and Powell went to great lengths to disseminate their ideas about water and to influence public opinion. In their

6 This "hydraulic policy" has been defined as "construction works planning". [Spanish Environment Ministry 1998, 701]

7 This "fin de siécle" crisis, as it been called, was caused by the expansion of international cereals markets in the second half of the 19th century driven by the revolution in transport and shipping, which had a tremendously adverse impact on the traditional pattern of cereals production and trade in the arid regions of Spain.

8 Common Law in Upper Aragon.

9 Agrarian Collectivism in Spain.

lifetimes this discourse was largely ignored, but it was soon picked up again in both Spain and the USA when politicians began to see the political value of large water infrastructure projects.

The biographies of Joaquín Costa and John W. Powell share a number of common features that make comparison significant. Moreover, the similarity of their ideas appears all the more significant in the terms in which they couched their arguments. By way of illustration, we may consider the following description of agricultural needs in the American West:

> About two-fifths of the area of the United States is so arid that agriculture is impossible without artificial irrigation, the rainfall being insufficient for the fertilization of ordinary crops. In this region all agriculture depends upon the use of running streams. In all of this country, wherever agriculture is prosecuted, dams must be constructed, and the waters spread upon the lands through the agency of canals. Again, as the season of growing crops is comparatively short, – in most of the country it lasts from two to three months only, – the waters of the non irrigable season will run to waste unless they are stored in reservoirs. Already the storing of these waters is begun; the people are constructing reservoirs, and will continue the process until all of the streams of the arid region are wholly utilized in this manner, so that no waste water runs to the sea (Powell 1889, 152).

In 1880, Joaquín Costa delivered a speech advocating a water policy at a farmers' congress in Madrid:

> The basic condition for social and agricultural progress in Spain today is bound up with the construction of pools and reservoirs of rain and running water. These works must be built by the nation and this Congress must urge our parliament and government to address the issue, asserting its place as the supreme aspiration of Spanish agriculture (Costa 1911, 4).

Just eight years separate these statements and their very proximity in time demands comparison. On both sides of the Atlantic, Powell and Costa expressed very similar ideas, though neither had any knowledge of the other. Both argued forcefully that the future of agriculture in an arid land depended on building large dams and reservoirs to store water and irrigate the fields in the dry seasons, the very time when crops could be grown. This was the basic principle underlying the "water" policy. The notion was not at all new, of course, and had been known and applied since antiquity. Both the Romans[10] and the Moors applied it in Spain, for example. Spaniards would introduce new hydraulic techniques in California in the 18th century. However the water policy went far beyond the mere application

10 The Proserpina dam in Spain was built by the Romans in the 2nd century close to the city of Emerita Augusta (the modern Mérida) and it is still functioning.

of the principles of hydraulic engineering. It implied a whole range of variables, which in my view were already present in the thinking of both Powell and Costa. The water policy was a program for national development, and it was endowed with the power to transform the agriculture of the entire nation. It is true that Powell's ideas referred strictly only to the arid West, but this is still a vast territory, while Costa's program focused on the arid parts of Spain, but that is most of the country. The water policy was thus national and not local. Nature existed to be transformed by the growing power of modern engineering. The belief in the power of science and technology inspired Powell's stricture, "so that no waste water runs to the sea." To consider that any water reaching the sea was a "waste" meant believing it was possible to dam and regulate all running waters for productive use. This extraordinary chimera, impossible to achieve and even undesirable according to today's environmental standards, nevertheless inspired gigantic construction projects in the form of large dams and water transfer infrastructure in both the USA and Spain[11]. Such an enterprise could only be achieved by the State, as Costa himself made clear. Powell also demanded Federal intervention to irrigate the arid lands of the West. The state and irrigation have been closely intertwined[12] ever since the days of the Mesopotamian civilizations, but with the rise of liberal capitalism in the second half of the 19th century it was no longer easy to maintain such ideas, and still less so in the United States.

By means of this threefold comparison I mean to stress how both Powell and Costa painted the same picture of the transformation of a territory and a society through the intensive exploitation of rivers. This endeavour goes far beyond the construction of large-scale dams and irrigation works, however, implying that water must be made the axis around which the whole of society, the economy, politics and culture must turn. In this light, I do not think that water policy in arid countries should be considered in historical terms simply as a series of government-inspired construction measures. Rather, it is the "grand" policy that shapes the "macropolitics of regimes" as Wildavsky puts it, and mobilizes a whole society in pursuit of wealth and happiness.

A New Landscape

What lies behind all of these proposals? What contexts should be considered if we are to understand Powell's and Costa's intentions? To begin with, historical circumstances explain some common features. In the second half of the 19th century, the USA was still engaged in the conquest of the West although the

---

11 As Marc Reisner remarks in the Introduction to *Cadillac Desert* (1987), "By the late 1970s, there were 1,251 major reservoirs in California, and every significant river – save one – had been dammed at least once."

12 Karl Wittfogel (1896–1988) was the author of the "Hydraulic Theory" of despotic states, which is based on the argument that the control and distribution of water had spawned authoritarian, centralized empires and sprawling bureaucracies (Wittfogel 1957).

situation was no longer comparable to the time of the pioneers with its trappers, caravans, Indian wars and gold fever. This was a second conquest in which only a few territories remained to be explored. The challenge now was to settle new populations in the vast and arid new territories. The colonization of the wilderness by farmers was, in the words of Wallace Stegner (1992), the "second opening of the West". This was the historical context of Powell's career and, crucially, of his personal experience of fieldwork among the Indians; exploration provided him with first-hand information and a vision of the landscape of the American West. He perceived the gulf between the arid lands of the new territories and the rest of the country, and he was keenly aware that the Jeffersonian ideal of a farmer cultivating God's garden in the form of 160 acres of green fields, which was the inspiration for America's farm policy at the time, made little sense in the West, where a farm of such a size without irrigation could not sustain a family, but 160 acres of irrigated land was really too much. In his introduction to Wallace Stegner's *Beyond the Hundredth Meridian,* Bernard De Voto writes:

> There is no need to describe how the "quarter-section" acquired mystical significance in American thinking – the idea that 160 acres were the ideal family-sized farm, the basis of a yeoman democracy, the buttress of our liberties, and the cornerstone of our economy. [...] But in the arid regions 160 acres were not a homestead. They were just a mathematical expression whose meanings in relation to agricultural settlement were disastrous (Stegner 1953, xix).

Powell's ideas for a new water policy were thus inspired by a fresh approach to the Jeffersonian ideal that would allow the fundamental tenets of the American Revolution to work in the arid West. Powell frequently complained that the US administration and politicians did not understand the nature of the West, viewing it from an Eastern standpoint, from the perspective of a more verdant land. However, he was also aware that it was his task to persuade the political establishment and bring public opinion around to his hydraulic ideas, and in 1878 he wrote a *Report on the Lands of the Arid Region* (1879), which he submitted to the US Congress.

In this work, Powell is something like a prophet, painting a panorama of the land to be colonized, an arid wilderness that generously offers wide open spaces, broad horizons, vast plains, canyons and snow-capped mountain ranges holding huge reserves of water. Powell's vision involved humanizing these enormous territories. This monumental task required building from scratch to dominate and profit from the wilderness, and this is surely the fundamental meaning of colonization. Intense hydraulic activity would be the main tool for this achievement. Costa took a surprisingly similar stance. His own historical context involved the profound crisis of the world of the Spanish peasantry and, more broadly, the struggles of the Spanish nation and the bankruptcy of a political system run by and for local political bosses known as *caciques*[13]. Costa was part of the "Regenerationist" movement that gathered force after the symbolic year 1898 when Spain lost Cuba, the Philippines and Puerto Rico, the last colonies of the old Spanish empire, following its defeat by the United States. These events represented the nadir of Spanish decadence for the Regenerationists, whose goal was to bring about the revival of the Spanish nation. Though he sympathized with this movement, Costa's work reveals a singularity akin to Powell's. He too held a vision of a landscape: the arid plains of his native Aragon and of Spain. Like John W. Powell, who contrasted the visions of the East Coast and the West of the United States, Costa compared arid Spain with the countryside of France. In 1867, the year of the Universal Exhibition, Costa spent nine months in Paris, and he returned with a vision, writing in his diaries that 1867 had been the year of his awakening. He was 21 years old at this time, and all of his biographers have remarked on the crucial importance of this experience for Costa's future intellectual development. The transformation of the dusty plains of Spain into a lush, green European countryside had become his dream, but it was a dream that he sought to make a reality through his hydraulic policy.

The early hydraulic discourse in both Spain and the USA was structured around the concepts of water, the landscape and the nation. The Jeffersonian Eden or "God's garden" and the green countryside of Europe were the examples that inspired the thought of both Powell and Costa as both a vision and a landscape, and as a part of a "national" ideal intended to achieve the revival of the homeland. In the USA, this revival was needed if the vast expanse of the new territories was to be colonized and assimilated, while in Spain it was a question of resurgence from an enervating decadence. Where the USA faced the challenge of applying the ideals of the American Revolution in the West (Powell's key idea), the challenge for Spain was Europeanization according to the Regenerationists. This aspiration produced much theorizing, but Costa was the only member of the movement who succeeded in constructing a discourse, harnessing his imagination and rhetorical skills to paint for his audience a picture of his ideal, a landscape of cows grazing in green meadows, fields golden with crops, patches of woodland, gently winding streams, boats plying the canals, steeples rising above the towns and villages, roads and railways. Costa used this attractive vision to draw his audience's attention to the basic ideas and principles embodied in his water policy, above and beyond the benefits inherent in the creation of agricultural and environmental utilities, suggesting new markets, credit and industry, and the attainment of personal freedom. His vision was intended to transmit a clear idea of a new policy and a new way of governing, and it was founded on the stark contrast between the hard, bare lands of Spain, where poverty and oppression dwelt, and the land as it could be, redeemed by water and the expansion of irrigation. Water not only transforms

13 Costa roundly condemned the Spanish politics of his time, which he referred to scathingly as "caciquismo", a system based on permanent electoral fraud and the regular appointment of corrupt representatives and, by extension, corrupt parliaments and governments (Costa 1901).

farming and the landscape; it also liberates society. This was Joaquín Costa's basic principle.

Joaquín Costa campaigned tirelessly to spread the word. In 1892, for example, he delivered a lengthy speech in support of irrigation works in the town of Tamarite de Litera in the Aragonese province of Huesca to an audience made up mainly of farmers, small landowners and farm labourers. In this speech he pleaded for the construction of a new canal that would draw water from the River Esera to irrigate the district of Litera. Referring to the river, he told his audience:

> The river will create everything for you: government, policy, order, freedom, industry, trade, agriculture, railways, roads, churches, hospitals, schools, factories, theatres [...] [*will flow from*] the waters of your creative river: for you, the conservatives, it will bring order; for you, the liberals and republicans, it will bring independence and liberty; for the poor, wealth; for the rich, opulence; for the town, abundant revenues, public fountains, sewerage, avenues and street lighting; for the priests, piety and virtue; for the teachers, consideration and respect; for the usurer, ruin; for the jailer, idleness; for the artisan, a workshop transformed into a factory; for the emigrants, the road back to their abandoned homes [...] and strength and riches for the resurrection of our poor Spanish fatherland, which will never again be great or take a seat among the gathering of the nations, or spread over the planet or play an active part in the making of contemporary history while it remains an arid land ... (Costa 1911, 123).

This is Costa at his most rhetorical. His vigorous discourse was never improvised but was always the result of careful thought and deep reflection. Costa's speeches are erudite, but he had the literary skills to lend them considerable strength. Furthermore, no-one knew the world of the peasant farmers and landless labourers better than he did, for he had been born into it. This personal experience allowed Costa to arouse the feelings and emotions of his audiences, while his literary talent led him to portray water as the panacea for all the ills of the nation, a way to allow all members of society to participate in the attainment of a common good. Costa's hydraulic policy is, then, a utopia, the illusion of a balanced society able to cater for all interests. The revitalized nation will be made up of free and independent individuals. This is the political principle that imbues the hydraulic thinking of both Costa and Powell. What is peculiar to these men, however, is how they based freedom on the availability of water and conceived the transformation of the landscape as representing the liberation of society from servitude, poverty and oppression.

A Gospel of Water

Utopias, chimeras, colossal projects and revivals are very often the product of grand visions. A series of ideas forming a doctrinal corpus needs to be given shape in writing, a difficult task that is not always successful and in any case places considerable demands upon the reader. A vision, however, can be expressed orally, in music, in pictures, photographs and film as a representation of a reality that does not exist and is yet imminent. The challenge is to make such yearnings seem real, or almost real. Christianity was born from the Biblical texts, which lent the new religion an authority that the oral transmission of the new religion had lacked. Yet it was necessary to represent the written word in pictures to spread the message more broadly, and it was this need that gave life to the whole tradition of Western religious art. Text, narratives, discourse and pictures may all structure the spread of ideas, but images are undoubtedly the most powerful vehicle to imbue any system of thought with visibility as an ontological quality. What is "seen" exists or at least could exist. A vision is a representation that the audience can visualize for itself. The problem, of course, is how to produce this effect. Joaquín Costa placed his narrative in the context of traditional tales with which his audience of small landowners and farm labourers were highly familiar, and which were therefore powerful vectors for the transmission of images. These tales were drawn from the Bible, especially from the Old Testament. The Egypt of the Pharaohs, the Nile, Exodus and the crossing of the desert, the Euphrates and the Tigris of ancient Mesopotamia and the Promised Land of Israel were the source of numerous metaphors used by Costa to convince his listeners and readers of his ideas for a new water policy. These visions referred continually to the transformation of the desert into a flourishing garden.

The image underlying the water policy propounded by both Powell and Costa envisages the territory as a landscape waiting to be created through irrigation. Beyond this vision, we may also discern a rich world of metaphor shared by both men. Though Powell's writing is less rhetorical, he, like Costa, uses the Bible as a source of imagery.

> The arid lands of the West, last to be redeemed by methods first discovered in civilization, are the best agricultural lands of the continent. Not only must these lands be redeemed because of the wants of the population of that country, they must be redeemed because they are our best lands. All this is demonstrated by the history of the far West, and is abundantly proved by the history of civilized agriculture. All of the nations of Egypt were fed by the bounty of one river. In the arid region of the United States are four great rivers like the Nile, and scores of lesser rivers, thousands of creeks, and millions of springs and artesian fountains, and all are to be utilized in the near future for the hosts of men who are repairing to those sunny lands (Powell 1890, 768).

The Nile and ancient Egypt are frequently mentioned in Costa's writings, along with other Biblical references:

> In antiquity, at the dawn of history, Egypt was a region of howling desert. Yet Nature poured the waters of the mighty Nile there, and that river was managed

> with skill and back-breaking work century upon century to transform the arid desert into a flourishing garden. [...] Here you have my thinking about Litera: it is a region to be created, and its Maker will be the River Esera (Costa 1911, 134).

Comparing Powell's and Costa's use of this example, we may see that their meaning and intent was similar. The Nile as the nursing father of the land of Egypt is a perfect illustration, especially in Costa's discourse, of how the arid lands can be transformed by water. There is a religious echo in expressions such as "last to be redeemed" in Powell or "it is a region to be created and its Maker will be the River Esera" in Costa.

The Mormons of Utah were long the model of success in greening the desert. Powell mentions their efforts in his *Report on the Lands of the Arid Regions*:

> In Utah Territory cooperative labour, under ecclesiastical organization, has been very successful. Outside of Utah there are but few instances where it has been tried; but at Greeley, in the State of Colorado, this system has been eminently successful (Powell 1879, 11).

The extreme aridity of the land inspires a "moral" vision, and any humanizing action therefore represents an extraordinary and praiseworthy effort. The symbolic apparatus of religion helps to emphasize and represent this giant task. The transformation of the desert into a bounteous garden is an old story told in its original form in the Book of Exodus, which describes how Moses led the Israelites across the desert to the Promised Land. Mesopotamia, the wealthy land between the Euphrates and the Tigris is also mentioned in the Bible as the homeland of Abraham. My contention is that both Powell and Costa consciously used these stories for their rhetorical and emotional effect.

In his book *Cadillac Desert,* Reisner calls the transformation of the West "messianic":

> Everything depends on the manipulation of water – on capturing it behind dams, storing it, and rerouting it in concrete rivers over distances of hundred of miles. Were it not for a century and a half of messianic effort toward that end, the West as we know it would not exist (Reisner 1987, 3).

The messianic[14] is a key part of the Judaic and Christian traditions, which figures frequently in history, and Powell, the son of a Methodist farmer and preacher, grew up in Illinois, imbued with Christian ideals. Worster's description

14 "The Messiah is the one who announces and ushers in the heavenly kingdom on earth; he is the one who brings redemption for a community ..." (Pereira de Queiroz 1978, 21).

of Powell's parents and family, who had emigrated to the USA from Wales, is intended to distance them from the Mormons:

> They had brought with them the true Christian faith, based on the traditional Bible, and they needed no other, certainly not one from an upstart bumpkin who said he had seen angels (Worster 2001, 17).

Yet as the new capitalism developed and grew in strength all over the USA, the times did not favour Christian ideals such as piety and poverty which so inspired the Methodist morality:

> The harsher truth was that Illinois, indeed the whole United States, was making it increasingly difficult for pious men and women like the Powells to keep their affections firmly trained on God (Worster 2001, 52).

John W. Powell thus experienced the contradictions of a time when the morality of a revitalized Christianity in search of purity clashed with a society that was mobilizing for rapid enrichment. To go west was good. Redemption had become an ideal, morally constituted indeed, but more secular than sacred. The nation and science could redeem the arid land and society thanks to a new geography of irrigation. The water, the landscape and the nation became the "trinity" of a modern gospel of water.

The Bible is often present in Costa's speeches as a source of rhetorical inspiration. After his death he was represented as a new Moses, the prophet of redemption for the land and society. His epitaph in the impressive mausoleum erected in his honour in the cemetery of Zaragoza in 1915 reads:

> Aragon to Joaquín Costa, new Moses of a Spain in exodus. With the rod of his impassioned words he brought forth the spring of water in the sterile desert. He conceived new laws to bring his people to a land of promise. He made no laws himself.

These words display a rich rhetoric representing Costa as a prophet, a man who would become a symbol for the identity of the Aragonese people in their struggle for new irrigation works after his death. He was to be transformed into the prophet of a new gospel of water. Ironically Costa was not a Catholic, or even a Christian, and when he died the Bishop of Zaragoza refused permission for him to be buried in the Catholic cemetery. Nevertheless, he was always a devoted student of the Bible, which he regarded as a seminal text "full of political teaching".

A New Redemption

Redemption is a central concept of Christianity, expressing the key event of the New Testament story: God became a man to save mankind. Salvation thus becomes the basic message of Christianity and is announced by the Messiah. The minimal narrative structure of the New Testament was allegorically shifted by Joaquín Costa towards the transformation of the land into irrigated fields in a context of crisis. The arid land is in its death throes. Water appears as the instrument of salvation, a prophet emerges like a new Messiah to announce the good news, and a great water discourse is born, the mythical rhetoric of which is inspired mainly by the Bible.

In any event, neither Powell nor Costa sought to found a new cult or to spread the message of any existing religion. Rather, their aim was to renew the nation through the transformation of the landscape, and we may therefore affirm that their objectives were secular.

> True to his parents' example, Powell was saturated with moral fervour. Like them, he aspired to transform the world into a better place. Instead of the Protestant doctrines of Wesleyan Methodism, however, he turned to those other great nineteenth-century gospels of salvation, the nation-state and natural science [...] Science was to be a means of redemption. That redemption should begin with the nation they loved, cleaning out the cobwebs of its past – the sectional rivalries, the outworn creeds, the destructive greed. Thus redeemed, America could lead the world toward enlightenment (Worster 2001, 437).

The narrative of promise formed the mythical underpinnings of water policy as conceived in the USA and in Spain in the early 20th century. Water would reach the desert, requiring colossal works of hydraulic engineering, and the word was announced like a gospel. Let us look at a historical example that was of enormous importance in Spain.

The *Ley del Plan de Riegos del Alto Aragón*[15] was passed by the Spanish parliament in 1915. The Act envisaged irrigation of 300,000 Ha, equal to 25 per cent of the total area under irrigation in Spain at the time. This plan has taken an extraordinarily long time to put into practice. Work began on the 29th of March 1915 and it was originally envisaged that it would be completed within 25 years, but in reality the end is not in sight even today[16]. The very wording of the Act included value judgments about nature itself for example, describing non-irrigated land as *mezquino*[17]. An excerpt from the stated purpose of the Act, reveals the rhetoric used to explain the meaning of the Plan as an instrument to transform both agriculture and society.

---

15 Upper Aragon Irrigation Act.

16 The 300,000 Ha envisaged in 1915 were reduced to 172,000 Ha in 1956 in order to take account of financial and technical factors, which made it necessary to establish quality criteria for the land to be irrigated. Today, the area under irrigation within the geographical scope of the Upper Aragon Irrigation Plan is around 120,000 Ha. Further expansion of this area has been a matter of intense debate in recent years, given concerns about the likely returns on investment.

17 Wretched.

> This implies, then, a real geographical rectification, which will everywhere transform the wretched dry land where agriculture is today impossible into a great oasis, relieving the miserable condition to which the land is condemned by a generally dry climate that allows only winter crops and meagre, insecure yields. The Aragonese steppe will become a garden. (Proyecto de Riegos del Alto Aragón, 24)[18]

The transformation of the Aragonese "steppe" into a "garden" would be taken as the "promise of water", a transmutation of the hydraulic discourse of the Biblical Promised Land as the mythical basis for the Water Policy. The ideas deployed in a series of mythical narratives would need the help of engineers to carry out the enormous infrastructure projects required to realize the imagined landscape described by the "prophets" of redemption. Let us consider the words of one of the most important Spanish engineers, Manuel Lorenzo Pardo[19] (1989–1951):

> … in this region, which lies at the foot of the greatest water collector (the Pyrenees) in the Peninsula to which the Ebro gives its name, there are wide lands that have waited long ages for water and will use it with redeeming success. These steppes, punished as they have been by implacable drought […] were where the heralds did their work and the apostles of economic redemption through water preached the word, awakening souls to conviction and hope and preparing the way for action (Díaz Marta 1997, 51).

In this short passage we find some of the basic building blocks of a narrative that runs through 20th century Spanish Water Policy: the "howling desert", "punishment", "antiquity", "redemption", the "oasis", the "garden", the "word", "preaching", "prophecy", the "apostle", "souls", the "promise" and "hope". Why were science, technology and economics associated with such religious rhetoric?

> In the New World, Indians had dabbled with irrigation, and the Spanish had improved their techniques, but the Mormons attacked the desert full-bore, flooded it, subverted its dreadful indifference – moralized it – until they had made a Mesopotamia in America between the valleys of the Green River and the middle Snake. 56 years after the first earth was turned beside City Creek, the Mormons had six million acres under full or partial irrigation in several states. In that year – 1902 – the United States government launched its own irrigation

18 Proyecto de Riegos del Alto Aragón. 1913. Barcelona: Tip. El Anuario de la Exportación. p. 24.

19 Manuel Lorenzo Pardo was a brilliant engineer who organized the Confederacion Hidrografica del Ebro, the first watershed management authority in the world. He went on to become a leading figure in the promotion of the National Hydraulic Works Plan of 1934 and Director General of Hydraulic Works in the last years of the Second Republic.

> program based on Mormon experience, guided by Mormon laws, run largely by Mormons (Reisner 1987, 2).

This predilection for religion in the formulation of the principles of the "grand idea" need hardly surprise us. In the USA, it was the Mormons who were first to irrigate the deserts of the West in Utah, firm in the conviction that they were transforming the "promised land" into a garden. As Marc Reisner explains in *Cadillac Desert*, irrigation had become a kind of "Christian ideal".

The Political Value of Water Policy

Aaron Wildavsky[20] discusses the political value of the Moses myth in his work *The Nursing Father. Moses as a Political Leader*, in which he examines the mythical figure of Moses and his importance in the construction of political leadership. A crucial point of his argument, which I would like to stress here, is that Moses the leader disappears in the book he is given by God for his people. The discourse is where the leader who communicates the "great promise"[21] remains, whether it is the Promised Land of the Jews, the new lands of the West for the Mormons, or the great garden into which the arid Spanish plain was to be transformed. In my opinion Water Policy has always been linked with founding myths wherever it has existed, whether in the USA, the Soviet Union or Spain. These myths were useful for spreading the idea, which could not otherwise have been justified, of transforming vast areas of land at such enormous economic and human cost. A transcendental message inspired by Christian and revolutionary ideals like the one delivered by Joaquín Costa in Spain was what was needed.

Water Policy has proved fertile ground for the construction of political leadership at different times and in different contexts. The rulers discovered the political value of Water Policy, and this political practice has lasted until the present day. In Spain the short-lived Second Republic (1931–1936) pressed ahead with the Water Policy from a theoretical standpoint, drawing up the First National Plan for Hydraulic Works (1934), and in terms of actual construction. Under the ministerial guidance of a socialist like Indalecio Prieto and thanks to the efforts of engineers like Manuel Lorenzo Pardo and Félix de los Ríos, the Republican period was one of brilliant success in matters of irrigation. So much so, indeed, that the

---

20 Wildavsky, A. 1984. *The Nursing Father. Moses as a Political Leader*. The University of Alabama Press.

21 The famous speech in which Martin Luther King promised a future of equality to thousands of his black followers in Washington was based on a prophetic vision which he described as a dream. "I have a dream" were the words King chose, and let us not forget that he was himself a Protestant pastor, to convey forcefully to his listeners that the promised future of equality was not one he expected ever to see. Shortly afterwards he was assassinated; it was then that his words became the foundation for his myth, into which he disappeared on his death.

Franco regime did little more than continue with the implementation of the same Hydraulic Plan. Between 1922 and 1930, the period of General Primo de Rivera's dictatorship, dams with a capacity of 514.5 Hm^3 were built. Between 1931 and 1936, the time of the Republic, this figure rose to 2,522 Hm^3, and between 1939 and 1966, in the midst of Franco's dictatorship, some 21,364 Hm^3 were completed (Díaz Marta 1998, 17). Despite the convulsions Spanish society suffered during these decades, the construction work that was the defining feature of the Hydraulic Policy slackened only during the Civil War, increasing continually until the latter years of Franco's rule. In my opinion, the Hydraulic Policy was able to transcend the fault lines in Spanish society because it alluded to the myth of the "promised land", which meant it could serve both the right and the left, offering redemption and the promise of salvation both for the sinner and for the oppressed. As a prophet of redemption through irrigation, Joaquín Costa would form a part of the pantheon of illustrious figures for both the Republicans (including the anarchists of the CNT[22]) and for the followers of Franco, who were happy to put up monuments in his honour and name streets and buildings after him.

It was Franco who was able to put the Hydraulic Policy to maximum use as a propaganda tool for his dictatorship. As early as April 1939, as the bloody Civil War drew to an end, the Nationalist regime promulgated a *Plan General de Obras Hidráulicas*[23] that continued the hydraulic planning approach established by Lorenzo Pardo in his Plan of 1933. The capacity of Spanish dams rose from 3,600 to 42,000 Hm^3 between 1940 and 1970, while the area under irrigation grew from 1.4 million Ha in 1940 to around 2.2 million 30 years later. As Gabriel Tortella writes:

> … the capacity for construction of an authoritarian state, which achieved perhaps its greatest economic success by applying the principles of a water policy developed in an earlier period, and the increased scope for investment in infrastructure that came with rising national income allowed the realization of this major programme of works[23] (Tortella 1994, 239).

The task of the anthropologist is to analyze and interpret culture, and it is for this reason that I propose a historical and cultural reading of Spanish Hydraulic Policy in the 20th century. The fact that the same policy could have driven the action of such different political regimes may, in my opinion, appear paradoxical if we do not take into account the mythical dimension and the rhetorical value of the policy as an instrument to make the myth visible as a horizon of promise. At this point we can appraise the symbolic value of the myth. The "promise of water" had the capacity, as Aaron Wildavsky shows, to produce and reproduce leadership in

22 The Christian myth of redemption appears to have formed part of the ideology of the Spanish Left from early on. This trait is perhaps most clearly apparent among the anarchists.

23 General Hydraulic Works Plan.

a traditional society like Spain's, which was struggling to achieve modernization and Europeanization but was a country in which the religious devotion of the past still had great power. I believe that the Spanish elites[24] have made of Water Policy, whether in the form of national plans or regional programmes, an instrument to win and wield power in a society where propaganda echoes increasingly loudly in the mass media right up to the present day. It is precisely the reactivation of this paradoxical water policy which reawakens myths and promises, that defines the debate over water in Spain today.

Water Policy and Democracy

Following the death of Franco in 1975, the construction of irrigation works stagnated throughout the period of the Democratic Transition, and the Hydraulic Plan was practically halted. Only a few projects that had already been set in motion in the latter years of the dictatorship were completed, and not without enormous difficulties. The Hydraulic Works Plan fell into disrepute, as it was directly associated with the former regime and with the image of Franco opening dams. Nevertheless, it is interesting to observe how the civil engineers in the service of the government, mainly those in the Watershed Authorities or in high positions in the Ministries, have tried to ensure the continuity of the same Water Policy, even in the midst of vociferous debate.

In 1976 a plan for the regulation of the River Esera (Huesca), intended to expand irrigation along the Aragon and Catalonia canal by building a huge dam to stock 600 Hm^3 that was to be named after Manuel Lorenzo Pardo gave rise to the first water protest movement. The movement was able to act in the comparative freedom of a fledgling democracy. This opposition was decisive in foregrounding the difficulties that a traditional Hydraulic Policy was likely to face in the modern democracy that was gradually taking shape in Spain (Mairal and Bergua 1997). This was followed by the cases of the Itoiz dams in Navarre (1985) and in Jánovas, Santaliestra, Yesa and Biscarrués in Huesca in the 1990s, which made even bigger headlines as examples of the social and political conflict caused by new schemes and the construction of new water infrastructure.

The official data of the *Confederación Hidrográfica del Ebro*[25] reveal the discontinuity of new construction for irrigation works after the end of the Franco regime. This example refers to Aragon, a key region (Autonomous Community) for the Spanish Water Policy. The situation of dams under construction in 2003 speaks for itself.

To understand and explain this sudden decline in the construction of large dams requires a complex analysis combining numerous different variables. Above all, however, I would point to the crisis of the Hydraulic Policy model in relation to

24 In my opinion, civil engineers, and especially those who have held high office in government or in the leading construction firms, are a key component of these elites.

25 Ebro River Watershed Authority.

Table 2.1 Dams built and under construction in Aragon since 1931

Political regime	Number of dams	Capacity in hm^3
Republic 1931–36	3	152
Franco dictatorship 1939–75	17	1,865
Democracy 1975–2003	4	171
Under construction 2003	3	181.35

Source: Own work based on data published by *Confederación Hidrográfica del Ebro*, 2003.

the reconfiguration of Spanish society as a modern parliamentary democracy. The procedure by which a major hydraulic infrastructure scheme is defined, beginning with initial public hearings, has become ever more complicated and difficult. The legal complications surrounding the whole process take up a great deal of energy at the same time as offering increasing opportunities for court action by interested parties. Effective judicial oversight of major water infrastructure works has sometimes halted works for long periods. The Itoiz reservoir in Navarre could not be filled to the planned level for several years as a consequence of a ruling by the Spanish Supreme Court, and the courts annulled the whole administrative procedure carried out to authorize the plan for the Santaliestra dam in the Pyrenean district of Ribagorza. The plan was finally dropped. The plan for the Jánovas dam in Upper Aragon was shelved when it failed to pass the mandatory environmental impact assessment procedure. All of this means that the traditional Hydraulic Policy, which consisted of a whole compendium of schemes all over Spain at the time of Franco's death in 1975, has faced new challenges since the Transition, which in the final analysis boil down to the need to adapt to the new democratic conditions. Large irrigation works are more likely to meet with opposition, and democracy provides activists with a variety of routes to express their arguments and mobilize.

What has been called the *comunidad de política hidráulica tradicional*[26] (Pérez Díaz, Mezo and Álvarez 1996, 41) was formed by "politicians, administrators, economists and engineers in the service of the government, irrigators and construction firms", a charmed circle that controlled Water Policy in Spain throughout the 20th century. In the 1990s, however, this community was overwhelmed as a series of new agents burst onto the scene. On the one hand, there was the Green Movement which was, however, largely unsuccessful in matters of water policy, perhaps because it had cut its teeth on the anti-nuclear struggle and on conservationist issues. Associations representing local people affected by large construction projects proved a different matter and were able to create a social movement and an associationist network that has produced its own thinking on the issues (Mairal 2003). In addition, the academic and scientific world produced

26 Traditional hydraulic policy community.

a multidisciplinary movement of experts in water management that was able to bring critical ideas into the public forum. This in turn gave rise to a movement that combined both expert knowledge and political activism to formulate an alternative model for water policy in the year 2000, which has been called the *New Water Culture.* The consolidation of the "Autonomous Communities"[27] and the devolution to them of powers over water management in some cases, or the demand for such powers in others has been a political development of enormous import. As a result, a national policy has become regionalized, leading to water "wars" between different Autonomous Communities (Aragon, Catalonia, Valencia, Murcia, Castile-La Mancha).

The 2001 National Water Plan or the Paradox of Water Policy in Spain

The turning point for the stagnation of the post-Franco Hydraulic Policy was the 1985 *Ley de Aguas*[28], which conferred enormous importance on hydraulic planning through two basic instruments: *Planes Hidrológicos de Cuenca*[29] and the *Plan Hidrológico Nacional*[30] The latter is key, as it is the only instrument capable of redrawing the traditional Hydraulic Policy along more modern lines defined, among other factors, by a wider scope, the consideration of a range of new variables, a multidisciplinary approach to expert knowledge and an opening to society through participation by the citizen. This was the objective discernible in the new water planning procedures proposed in the 1985 Waters Act. Events since then have been controversial, however. While it is true that water has become a key political issue in Spain, it is no less true that the political agents who traditionally influenced the course of water policy have seen their position strengthened, because the issue provides a powerful instrument for mobilizing the population and winning or holding onto power. Let us consider some of the key events of the last two decades.

The first attempt by a Socialist government to formulate and implement a National Water Plan in 1993 also saw the first court cases on water issues, which would become key in all of the subsequent debates. In Spain, it is argued, some river basins have surplus flows that "drain away" into the sea, while others suffer water shortages. Hydraulic planning should seek to correct these imbalances. The

27 Regional political autonomy was introduced in Spain after Franco's death, and this form of devolution has resulted in practice in a federal-type state in which the Autonomous Communities enjoy wide powers. The exclusive powers of the central government over shared river basins are now being challenged by some of the Autonomous Communities, which want exclusive powers to manage their own rivers, even where these flow through more than one region. If this demand were granted, it would break the principle, enshrined in Spain in the 1920s, that the watershed is the unit of management.

28 Waters Act.

29 Watershed Plans.

30 National Water Plan.

only way to achieve this balance is to transfer water from those river systems with "surpluses" to those in "need"[31]. These arguments were used to justify the construction of further inter-river transfer infrastructures, and this decision would become the central issue of water policy in Spain at the turn of the 21st century, generating intense social and political conflict.

In 1996 the conservative *Partido Popular*[32] led by José María Aznar won power and began to revise all of the existing water policy. In the year 2000, the PP won an absolute majority in the Spanish parliament and the government embarked upon the enormous task of drawing up a new National Water Plan that would be much wider in scope than its predecessors. This Plan was enshrined in law in 2001 and implementation began immediately. However, it was the Socialists who won the 2004 elections, and one of their first measures was to revoke the PP's National Water Plan and to cancel the Ebro transfer, the main infrastructure project provided for. These events clearly show how hydraulic planning in a modern democracy can only take place in a stable context of political consensus, because the capacity to bring about structural change or legislate for the future depends on broad agreement as to goals and outcomes. The 2001 National Water Plan was not drafted with any such consensus in mind but imposed by a parliamentary majority that finally proved ephemeral, as is only natural in a democracy. This led ineluctably to the failure of the Plan. Meanwhile, the 2001 legislation failed to envisage any framework of possible political consensus for a hypothetical National Water Plan, either as a technical requirement or as part of any scientific analysis, thus missing the opportunity to design acceptable alternatives for multiple scenarios. Instead, the 2001 Plan, which contained some positive features, focused on the project to divert an annual 1,050 Hm^3 from the River Ebro to part of Catalonia, Valencia, Murcia and Almería. This provoked an intense backlash of protest in both Aragon and Catalonia. The transfer was only a part, though a very important one, of the 2001 National Water Plan, but it galvanized the interest of politicians, experts and public opinion, making water into a key issue nationwide and sparking a fierce debate unparalleled in any other country.

In 2000 the European Union Framework Directive on Water became the basic instrument for defining what might be a new water policy. However, water policy in Spain is in the throes of a difficult and costly transition from the 20th century *Política Hidráulica* as a Water Policy model to a new, more complex, multidisciplinary model that has still not achieved the level of realization or the legislative provision[33] of its predecessor.

---

31 Criticism of these principles has become a key issue in the debate over water policy in Spain since the 1980s. One major transfer already exists in Spain. This is the Tagus-Segura diversion, which was built in the 1970s.

32 Popular Party.

33 A new proposal for legislation has finally been put on the table, which is none other than that outlined in the "Nueva Cultura del Agua" manifesto proposed by the eponymous Foundation. This proposal for a new water policy has gained considerable support, and the

In my opinion, Water Policy in Spain has been rather a means of practising politics than a policy as such, and it is because of this that it has such had such a profound influence. It has resurfaced as a way of achieving and a style of exercising power, as can be seen in the Autonomous Communities of Aragon, Valencia, Murcia and Castile – La Mancha. In this region, water is not presented as a resource, a public good or an environmental asset, but emotively, as a grievance linked to the local identity of a population that hopes to see the fulfilment of the "grand promise." The whole traditional rhetoric of the gospel of water that accompanied the old "Hydraulic Policy" has been revisited by contemporary politicians. In September 2000 a huge demonstration was held in the Aragonese capital of Zaragoza (according to the headlines some 400,000 people took to the streets, although this figure may be exaggerated), in which the whole gamut of emotions was reflected by a coterie of political leaders who led the march under a banner bearing the campaign slogan, *Aragón, agua y futuro*[34]. These words are as rhetorical as the context in which they were used and were intended to reflect Aragon's life-and-death dependence on water. Aragon indeed depends on water, but no more so than any other society or territory. In 2004 the Socialist government repealed the National Hydraulic Plan Act of 2001, in particular those parts referring to the Ebro transfer, which had aroused fierce protest in Aragon but also widespread support in Valencia and Murcia, not to mention heated debate throughout Spain. The discourse of the demand for and use of water (in this case the transfer) to mobilize the population was voiced on this occasion by Valencia and Murcia, both Autonomous Communities governed by conservatives, whose political leaders once again fell back on the notion of redemption and the thirsty land, suggesting catastrophic consequences for their regions if the Plan were repealed. Their constant tirades allude to a doubtful future and decry the lack of solidarity of those who have water to spare. The slogan this time was *Agua para todos*[35].

In all of these cases, water ceases to be a physical resource with multiple uses that can be stored and distributed, and that serves environmental and even recreational needs; instead it becomes a symbol of desire, concentrating feeling and unleashing intense emotions (Mairal 2003). When this world of value judgments and symbols confronts hydraulic planning and becomes a part of hydraulic policies instigated and actually implemented by government, the great paradox emerges that water is not simply H_2O but many things serving many different interests.

The transition in Spain from the traditional Hydraulic Policy, which was based on enormous engineering works, to a modern Water Policy based on multidisciplinary management of the resource, so necessary in today's world, is

Spanish Ministry of Environment appears not uninterested. However, it remains to be seen how matters will progress in the immediate future, when the ideas contained in the proposal will have to be defined in more detail.

34 Aragon, water and a future.

35 Water for all.

constantly skewed by the rhetoric of myth deployed by the political leaders of the Autonomous Communities in the context of the battle for power in their regions, and by the central government as a means of managing regional powers in their own partisan interests. In this light, we can only conclude that water policy in Spain remains a captive of historical forces.

References

Arrojo, P. and Naredo, J.M. 1997. *La gestión del agua en España y California*. Bilbao: Bakeaz.

Cernea, M. (ed.) 1991. *Putting People First*. New York: Oxford University Press.

Costa, J. 2005. *Política hidráulica: Misión social de los riegos en España*. Pamplona: Anacleta.

Cheyne, G. 1972. *Joaquín Costa, el gran desconocido*. Barcelona: Ariel.

— 1972. *A Bibliographical Study of the Writings of Joaquín Costa (1846–1911)*. London: Thames Collection.

Díaz Marta, M. 1997. *Las obras hidráulicas en España* Aranjuez: Ediciones doce calles.

Fernández Clemente, E. 1989. *Ensayos sobre Joaquín Costa* Zaragoza: Prensas Universitarias de Zaragoza.

— 2000. *De la utopía de Joaquín Costa a la intervención del Estado: un siglo de obras hidráulicas en España*. Zaragoza: Cuadernos de Escuela y Despensa Facultad de Ciencias Económicas y Empresariales.

Fiege, M. 1999. *Irrigated Eden. The Making of an Agricultural Landscape in the American West* Seattle: University of Washington Press.

Mairal, G. 2001. Discursos de riesgo y agonía. In Lisón Tolosana, C. (ed.) *Antropología: horizontes emotivos*. Granada: Editorial de la Universidad de Granada.

— 2000. Joaquín Costa y sus mundos. In: Del Campo, S. (ed.) *Historia de la Sociología española*. Barcelona: Ariel.

— 2003. The invention of a "Minority": A case study from the Aragonese Pyrenees. In: Boholm, Å. and Löfstedt, R. (eds), *Facility Siting. Risk, Power and Identity in Land Use Planning*. London: Earthscan Publications: 144–160.

Mairal, G. 2007. Las paradojas de la política del agua en España. *Panorama Social*. FUNCAS: 102–115.

Ministerio de Medio Ambiente. 1998. *El Libro Blanco del Agua*. Madrid: Ministerio de Medio Ambiente.

Pereira de Queiroz, María I. 1978. *Historia y etnología de los movimientos mesiánicos*. México: Siglo XXI.

Pérez Díaz, V., Mezo, J. and Alvarez-Miranda, B. 1996. *Política y economía del agua en España*. Madrid: Círculo de Empresarios.

Powell, J.W. 1879. *Report on the Land of the Arid Regions*. Washington: United States Congress.

— 1889. The lesson of Conemaugh. *The North American Review* 149: 393.
— 1890. The irrigable lands of the arid region. *The Century* 39: 5.
Proyecto de Riegos del Alto Aragón. 1913. Barcelona: Tip. El Anuario de la Exportación.
Reisner, M. 1986. *Cadillac Desert. The America West and its Disappearing Water* New York: Penguin Books.
Stegner, W. 1953. *Beyond the Hundreth Meridian. John Wesley Powell and the Second Opening of the West*. New York: Penguin Books.
Tortella, G. 1994. *El desarrollo de la España contemporánea. Historia económica de los siglos XIX y XX.* Madrid: Alianza Editorial.
Westerman, F. 2005. *Igenieros del alma.* Madrid: Siruela.
Wildavsky, A. 1984. *The Nursing Father. Moses as a Political Leader.* Tuscaloosa, AL: The University of Alabama Press.
Wittfogel, K. 1966. *Despotismo oriental: estudio comparativo del poder totalitario.* Madrid: Guadarrama.
World Commission on Dams 2000. *Dams and Development: A New Framework for Decision-Making.* Washington: World Bank.
Worster, D. 1985. *Rivers of Empire. Water, Aridity and the Growth of the American West.* New York: Oxford University Press.
— 2001. *A River Running West. The Life of John Wesley Powell.* New York: Oxford University Press.

Chapter 3
Water, Culture, and Civilization in the Italian Mediterranean

Eriberto Eulisse

Nullius fons non sacer.[1]

Servius

Water, as with all things considered only for their usage,
has become alienated from its own history and made invisible.

Renzo Franzin (2005)

In the history of the Mediterranean as in the rest of human history, major water civilizations and small traditional societies flourished and survived as a result of careful water management. Indeed the sustainable use of water is a necessary premise for the birth and the prosperity of all civilizations. In Mediterranean Italy too, remarkable and irreplaceable water cultures have arisen from the unique yet fragile relationship that its populations established with their own environment.

As in many other parts of the world, the populations of Mediterranean Italy have long invested water with intrinsic qualities, with sacred and life-giving connotations. Purifying waters, therapeutic springs that burst from the bowels of the earth, waters of fertility, sacred and oracular fountains, deified rivers: all these conceptions have their roots in the dawn of Mediterranean civilization.

Uses and customs pertaining to water, a fundamental element of life itself, are reflected in the cosmogonies, myths and rites of all Italic cultures. The aim of such symbolic and religious systems has always been to preserve this invaluable resource for the survival of the contemporary society and for future generations. When abuse or non-sustainable usage prevailed and the equilibrium of the ecosystem was violated, systems of production were irreparably damaged and societies fell into collapse.

Many written records, myths and legends of antiquity that have survived to the present constitute direct and vivid evidence of this attitude to water. In Greek Sicily, for example, the oracle of Apollo forbad any tampering with the Camarina marshes; otherwise a great calamity would strike the colonial city.[2] Examples like this from ancient Mediterranean Italy are common, as the following pages will make clear.

1 'All springs are sacred.' Servius, *In Vergilii Aeneide*, VII.84.

2 Vergilius, *Aeneide*, III.701.

Nevertheless, today's consumerist and information-based society has apparently dismissed similar warnings. Bombarded as we are each day by news from the mass media, it is truly surprising to observe just how little attention is given today to the delicate issue of water and its value. Indeed to speak of the 'value' of water is not to speak of its 'price' or cost.[3] We delude ourselves that an unlimited quantity of water is available for us, but just what is left of its sacred, or at least ethical dimension?

The purpose of this chapter is to illustrate the myriad concepts and perceptions associated with water in Italy from ancient times to the present day. Naturally, given the available space, only a partial and incomplete picture may be traced in the attempt to reconstruct the sense of awe and sacredness that water once invoked in specific cultural contexts. To refer to the contemporary dimension inevitably means to reflect upon the commercial logic that has erased the intrinsic value of this precious life blood.

Too often in the 'developed' world, the value of water is reduced to a mere economic question. As I will highlight with regard to Mediterranean Italy, this is also due to the technocratic paradigm of a society which has systematically ignored the aquatic ecosystems at least from World War II on, instilling a view of development that considers nature no more than a 'resource' to be exploited. If it is true that modern hydraulic technology has improved and enhanced our living conditions, it is also clear that the short-sighted features of consumer societies and its models of development have negative short-term and long-term consequences. In the pages that follow, I will present a perspective that provides keys for understanding the richness and complexity of those water cultures that existed in the ancient Italic world. By 'water culture', I refer both to non-material features (myths, rituals, and behaviours) and to important material advances associated with water, in the form of hydraulic engineering, and other archaeological and architectural evidence.

From this perspective I will examine the prehistoric culture of Sardinia, the Etruscan world, the Greek and Roman periods and last but not least, the folklore of Calabria. The choice of what to investigate is naturally determined by the available archaeological, iconographic, literary, and ethnographic sources. Much will be left in darkness because of the lack of written or material records. For the Greek and Roman periods, however, it is easier to reconstruct an adequate picture.

To reconstruct a picture of past water cultures, a combined historical and anthropological approach is necessary. In order to understand how water was so powerful in shaping identity and how it demarcated the diverse worlds in which real men and women lived out their lives, we must obviously refer to many different histories of water (Sardinian, Etruscan, Roman, etc), and not to a hypothetical homogeneous and unifying 'Italian water culture.' Though there are clear common features among these histories, water perceptions and conceptions

3 As Oscar Wilde noted in his *Lady Windermere's Fan*, 'a cynic is a man who knows the price of everything but the value of nothing.'

have to be understood above all with reference to the specific cultural context in which they were conceived.

From this methodological perspective, it is possible to reconstruct a whole civilisation on the basis of how water was considered and used according to its religious, symbolic, and artistic peculiarities. Indeed water constitutes a privileged base from which to reconstruct the identity and self-representational forms of any population and culture.

Any attempt to trace the history of those water cultures that have succeeded one another in Mediterranean Italy makes sense only if it leads to a better understanding of the present. To this end, the final section of this essay examines contemporary perceptions and 'myths' associated with water, from the modern myth of the presumed 'purity' of bottled water to that of infallible and all-powerful water technology. Similar perceptions and myths,[4] as I shall highlight, have led to widespread abuse and wastage, turning rivers into mere dumping grounds, leading to the depletion of resources and widespread pollution of underground water, and legitimizing development models that are completely short-sighted. Any vision of a sustainable future calls into question such models and paradigms of development.

Growing attention to the issue of 'traditional' knowledge and know-how stems from the conviction that this type of knowledge can be a key instrument in sustainable development for a better future (UNCCD 2005). In this sense, the traditional knowledge and techniques of ancient Italian cultures that I examine here should not be considered only as archival curiosities to be contrasted with modern technologies; rather, ancient and local solutions may be an exceptionally useful source of information that can help solve present problems such as flooding, desertification, and water scarcity. If we look more closely, the very problems we face today were tackled with far more wisdom and foresight in ancient times.

In rediscovering the traditional knowledge of the Mediterranean, it is possible to reinterpret it. Thus, recalling such knowledge is not simply a scholarly exercise, since many ancient expressions and conceptions embody complex longer timescales aimed at preserving resources for future use. Discussion of the critical issue of changing the present-day water perception cannot be postponed. The institutional aim of *Civiltà dell'Acqua* is to oppose a new Water Culture to the contemporary reductive vision of water and bestow it once again with the ethical values to which it is entitled and which were attributed to it from the dawn of humanity.[5]

4 In this chapter I shall refer to the word 'myth' as defined by Roland Barthes (1957) in his *Mythologies*.

5 The Water Civilization International Center (Centro Internazionale Civiltà dell'Acqua) is a non-profit association established in 1996 and based near Venice, Italy. Its activities are aimed to increase public awareness of the intrinsic value of water – a value other than its cost or price. The Center is active in fighting against present abuses and non

The Dawn of Mediterranean Civilization: The Sacred Wells of Sardinia

In prehistoric times, water had ritual and sacred connotations far beyond its economic and functional value. This is amply confirmed by numerous archaeological findings in various parts of the Italian peninsula. Exceptional among these remains are those found in Sardinia, an island located at the centre of important navigational routes in the Mediterranean since earliest times.

The sacred nature of underground water sources is confirmed in Sardinia from at least 1500 BC. Proof that the relationship between human society and water was permeated with religious significance can be deduced from the numerous underground water structures discovered on the island. These wells (or cisterns) with their characteristic dome-shaped covering are the so called 'sacred wells' or 'temple wells' of Sardinia, that are reached by an underground stairway and are also found in other parts of the Mediterranean. The archaeological sites of Romanzesu[6] (ca 1200 BC) and Santa Cristina of Paulilatino[7] (ca 1000 BC) confirm the highly symbolic significance of chthonic water sources in proto-historic Sardinia. Here, wells have been found in places not only distinguished by their proximity to underground sources, but also for their astronomical orientation (Lilliu 1982).

It has been observed, for example, that the mouth of the temple well of Paulilatino reflects the moon at its maximum declination, a phenomenon which occurs every 18 and-a-half years (the so-called lunar golden number). Furthermore the entrance lets the sunlight into the very bottom of the underground stairway only during vernal and autumnal equinoxes. The same characteristic lunar reflection is found at Funtana Coberta of Ballao and Su Puzzu of Orroli. Here, however, the stairway lets the sunlight in only at the summer solstice. These astronomical findings have led researchers to hypothesize that a fertility cult may have existed, relating the moon and stars to underground water (Rassu 2004, Altamore 2008).

These temple wells, 35 of which have survived to date in Sardinia alone, were constructed during the transitional period from the Bronze Age to the Iron Age. Archaeologists confirm that this period witnessed a veritable 'cultural revolution', both in material culture and architecture. Sardinia's sacred wells are among the oldest archaeological finds in Italy where some form of ritual observance connected to water and the chthonic world is adduced.

It has been argued that the typology of these temple wells in Sardinia was strongly influenced by Mycenaean culture. Indeed, at least 30 examples of underground wells with dome-shaped coverings with internal diameter varying from 3.5 to 5.5 meters (analogous to the Sardinian wells) exist in the Aegean. The underground fountains/cisterns of Myrtos-Pyrgos, Zacros and Archanes, all datable to between 1450 and 1100 BC (thus earlier than those in Sardinia) are

---

sustainable uses and promoting projects aimed at rediscovering the non material values associated with water in different cultures. See *www.civiltacqua.org*.

6 Municipality of Bitti, Nuoro Province.

7 Municipality of Oristano.

excellent examples of such structures. A particularly outstanding example is that of the Fountain of Perseia at Mycenae. The well-cisterns of Athens and Corinth also demonstrate similarities to the sacred wells of Sardinia, both in the underground stairway (composed of 99 steps) and the characteristic domed covering. Important parallels have been found as far away as Bulgaria, at a site about 200 km north of the Thracian Coast and 50 km west of Sophia. The site of Gârlo[8] shows many surprising morphological affinities with the Sardinian sites of Ballao and Santa Cristina of Paulilatino. These include the number of steps, the size of the dome, the depth of the well and its overall dimensions, which differ by only a few centimetres (Rassu 2004, Altamore 2008).

Taken together, the archaeological evidence suggests that a much broader area of the Mediterranean was involved in intense cultural and material contact and exchange than hitherto supposed. Water, with all its symbolic and sacred connotations, would not appear to have been excluded from this exchange, nor can we overlook the possibility that similar beliefs and rituals related to underground water may have existed in parallel with these common architectural forms.

Etruscan Masters of Water: Knowledge, Places and Cults

The Etruscans were long considered an obscure and mysterious people not just because their archaeological remains are relatively insignificant and scarce, but because none of their literature survives to the present day. However, over the last few decades a number of 'minor' archaeological discoveries have permitted a re-evaluation of their knowledge as applied, in particular, to hydraulic engineering.

Unlike the Greeks and Phoenicians who settled only in the coastal areas in Italy, the Etruscans also moved inland. Much of their attention was focused on the recycling and improvement of the soil. Their advanced agricultural techniques included the full use of rain water: in Etruscan territory, the abundant winter rainfall was carefully collected and stored for later use in the fields during the summer. Many examples of these very resourceful water systems, including the systematic harvesting of rain water, have been found recently throughout the Maremma.[9] The Etruscans established an advanced and well-planned economy in Italy, which, as several ancient sources testify, was envied even by Romans.

For the first time in European history, Etruscan agronomists applied techniques passed down from the thousand-year-old agronomical knowledge of Egypt and Mesopotamia. They were perfectly familiar with the construction techniques of canals and dams to make arid zones fertile, the art of drainage, and the construction of underground drainage tunnels (the Middle Eastern *qanat*). In recent years many discoveries of the evidence of this somewhat arduous work of land transformation have been made, thanks in part to the reclaiming of a number of territories after the Second World War.

8 Municipality of Breznik.

9 A coastal region between Tuscany and Lazio.

The techniques used for agriculture were also beneficial for urban areas. Indeed, each Etruscan city had a sizeable network of canals, terracotta piping, and artificial fountains to supply its inhabitants with drinking water. Very deep wells were dug which both channelled river water into the city and removed damp from the subsoil. Many of these hydraulic systems, which still form a labyrinth in the subsoil, survived in perfect working condition for more than 2,500 years. They were devised to keep the land drained and dry, and to channel the moisture into deeper strata. Similar hydraulic systems have been found wherever there were Etruscan villages and cities, for example, at Vetulonia, Bomarzo, Tarquinia and Chiusi. At Chiusi we find the famous Labyrinth of Porsenna, constructed between the 6th and 5th centuries BC, a veritable world of tunnels of all sizes on several levels below the city itself. The Labyrinth can be visited today. Legend has it that the Etruscan king Porsenna was buried in this inextricable underground labyrinth along with his treasure. Recent excavations have confirmed that what we have is an ancient and quite ingenious hydraulic system featuring both drainage and water conservation.

The Etruscan water structures may justly be compared to the famous Roman aqueducts which are a source of admiration to this day. It is well known that during the reign of Tarquinio the Haughty (descendant of the Etruscan dynasty of the Tarquinians)[10] Rome equipped itself with one of the most advanced systems of sewage and of waste-water collection in antiquity: the *cloaca maxima*. It was a system which Pliny went as far as to compare, among the wonders of the world, with the Egyptian pyramids (judging it to be, in comparison, both more useful and functional).[11] Despite these proven examples of hydraulic knowledge, many researchers have bemoaned the lack of a systematic study that fully illustrates the striking accomplishments of the Etruscans in the area of water (Vegetti and Manuli 1989).

As for the religious and symbolic aspects of water, archaeological, literary and iconographic sources enable us to reconstruct a no less interesting, albeit incomplete, picture. The sacred nature attributed to the Etruscan water sources is documented by the consistency with which votive offerings have been found both near underground water and hot springs, and in caves containing stalactites and stalagmites. The consistent number of *ex voto* objects found irrefutably confirms, according to Paolucci (2003), that the life-giving properties of these *fontes salutares*[12] were well known in Etruscan times – if not earlier. Water, in the Etruscan cosmos, is also clearly imbued with sacred connotations revolving around health, fertility and chthonic elements. The evidence of cults and sanctuaries dedicated to water which I will now consider is a clear demonstration of this.

10 Lucio Tarquinio, also called the Haughty, was the seventh and last king of Rome, before the Republic was proclaimed. He reigned from 535 to 510 BC, the year in which he was banished from Rome.

11 Pliny, *Naturalis Historia*, XXXVI.104–6.

12 Healthy springs.

Several reconstructions based on archaeological and iconographic sources seem to confirm that fresh water and its life-giving properties were presided over by some of the most important gods of the Etruscan pantheon: Aplu (the Etruscan Apollo), Vei (Demeter-Persephone), Uni (Juno), Diana Trivia (Phoebe) and Hercle (Hercules). In the Etruscan-Italic world, the demigod Hercules is often depicted either beside an amphora or about to fill one, or is in the proximity of rivers and fountains. This connection between Hercules and water is also to be found in the iconography of Greece and Magna Graecia, which I will analyse in the next section.

However, presiding over the Etruscan fresh water cult sites we find often a series of divinities who are superior to semi-divine heroes. Aplu is referred to at hot-spring sites such as Volterra and Chianciano.[13] Also, obvious fertility and chthonic connotations characterize the springs dedicated to Vei and Uni. Tibullo himself, in his *Elegies*,[14] links Etruscan water sites to the underworld and the mysteries of Persephone (Gilotta 2003).

At Vulci, in the sanctuary of Fontanile di Legnisina constructed between a series of caves and the River Fiora, a monumental Etruscan altar has been found next to a spring. Evidence of a local cult with significant chthonic connotations was deduced from an adjacent deposit of votive statues. An Etruscan inscription dedicated to the goddess Uni (here called 'protector of the springs') was found on one such object. Uni and Vei, who in many aspects preside over the agrarian-funerary sphere (thus fertility and regeneration), are the gods to whom this local cult was most probably dedicated (Ricciardi 2003).

Ritual forms connected to springs that originate in rocks and caves, and are also associated with the phenomena of stalactites and stalagmites, further confirm the strong link between underground water and the chthonic world that Etruscans believed in. On Mount Cetona, in the famous *grotta lattaia*[15] there is evidence of worship in the form of *ex voto* with statuettes depicting swaddled infants or seated youths. According to Paolucci and Manconi (2003), these finds support the existence of a water cult dedicated to birth divinities.

Near Volterra, an open air or 'natural' sanctuary was discovered (originally devoid of monumental remains) which is clearly linked to the chthonic world. In the sanctuary, several bronze votive artefacts came to light in areas that precisely coincide with the map of the springs. The spa complex of Sasso Pisano (between Volterra and Populonia), a place renowned for its intense geothermal activity, is also strongly associated with a chthonic health cult (Vitali et al. 2003).

From an iconographic point of view, the repertory of images pertaining to water that have been found on Etruscan gemstones, mirrors, sarcophagi, urns, bronze or terracotta vases confirms that this precious element was conceived both as a life-giving force and yet also imbued with chthonian aspects linked to the afterlife.

13 Siena Province.

14 Tibullo, *Elegies*, III. 5.

15 'Milkman's cave.'

Among the images that have survived, it is interesting to note that rivers (and the sea) are often represented as the means by which the passage to the underworld is reached: the depiction of the River Acheronte in the Tomb of the *Demoni Azzurri* (the 'Blue Devils') at Tarquinia (5th–6th century BC), is a good example (Gilotta 2003). Early literary sources also point to another important aspect of Etruscan springs, and that is their intrinsic and constant link to a tutelary or eponymous god. Tibullo, in his *Elegies*, states that wherever water springs forth in Etruscan territory there is always a connection with some local tutelary divinity. According to such a belief, rivers may be considered the personification of divine beings. Thus, for example, the River Chiana (which flows through three Etruscan cities: Chiusi, Cortona and Arezzo) seems to be linked to the eponymous god Klanins.[16] The discovery of statuettes with votive inscriptions dedicated to this divinity would confirm such an Etruscan custom (Maggiani 2003).

Here it must also be noted that the identity of these local divinities in association with springs and underground water, is not well defined. If it is true that Tibullo speaks of the 'gods of the Etruscan lifeblood',[17] as if they were real entities or forces, their specific names have rarely been passed down in written records. Thus, researchers tend to agree that these divine forces should be considered as a sort of *genius loci*, typical of Etruscan (probably also pre-Etruscan) religious beliefs. At the same time, according to Gilotta (2003), we cannot exclude the possibility that these manifold divinities associated with water,[18] albeit without precise names, were progressively assimilated by major divinities, such as Persephone, Demeter or Diana Trivia.

The wealth of archaeological remains associated with water sources in Etruscan territory led to the creation of a Museo delle Acque in 1997.[19] This museum, situated near the famous town of Chianciano Terme, contains a striking collection of material evidence for cults connected to hot springs and health-giving water sources: bronze and terracotta vessels made to contain water, amphorae (*loutrophoros*) for ritual washing, and *hydria* (water receptacles for everyday use). On display are also the considerable remains of the pediments of an Etruscan temple, erected in the vicinity of the hot spring *I Fucoli*, which Roman records confirm as being one of the most frequented hot spring sites in antiquity (Paolucci 2003). The springs of Fucoli are still active today and visited by numerous people for their diuretic and digestive properties.

Another water sanctuary discovered at Chianciano near the hot spring of Sillene, a little further uphill, endorses the strong link this area had with cults of health and water and, in this case, also with the chthonic world. The sanctuary of Sillene, founded at the end of the 6th century BC, is characterized by a cult related

16 Water name 'Klani'; Latin 'Clanins'.

17 *Tuscae numina lymphae*, III. 5.

18 Depictions of undefined divinities associated with water can be found on a number of Etruscan mirrors from the 3rd century BC.

19 Water Museum.

to the hot springs and the moon, as testified by the numerous inscriptions found here on many anatomic *ex voto* (casts of body parts: heads, eyes and legs). The magnificent bronze statues also found here do not allow for a precise attribution to any particular divinity. However, from the evidence, the sanctuary would seem to have been dedicated to Aplu or to Diana Trivia, goddess of the woods and hunting as well as of the moon and the afterlife. A more elaborate use of the water resources of Sillene and Fucoli, taking the form of imposing spa complexes constructed on the site, may date only to the Roman period (Maggiani 2003).

Myth and the Sacredness of Water in Greek Sicily

Greek Sicily benefited hugely from the influence of ancient Greece in terms of the knowledge and techniques of hydraulic engineering. Much of the Greek knowledge of hydrology was derived, in turn, from Near Eastern civilizations. Techniques for the construction of drainage tunnels,[20] hydraulic instruments and the practical application of hydraulic principles (such as siphons, water bellows and water clocks), all of which had been used for centuries in Egypt, were adopted by the Greeks, who both perfected them and extended them to the rest of the ancient Mediterranean world.

On the other hand, Greek Sicily's own contributions should not be overlooked, particularly those of Archimedes of Syracuse. To the Greek scientist, who lived in the 2nd century BC at the time of Syracuse's political and military expansion, we owe the earliest formulations of the principles of buoyancy. Archimedes has also been credited with the invention of the famous 'spiral pump' for drawing water, the use of which was widely diffused in the ancient world.[21]

In examining the conceptions inherent to the nature of water in Greek Sicily,[22] we must inevitably refer to the broader culture of ancient Greece. Elegies to water echo throughout Greek literature. Tragic poets, and Aeschylus first among them, invoked the gleaming, glittering water as the source of life. In the ancient Greek world, water sources are often personified, as in the case of Calliroe, the semi-divine heroine whose name literally means 'good flow.' The choice of the goddess's name is certainly not a casual one since water in ancient Greece was

20 Tunnels known as *qanat* in the Near East, and *krenai* in ancient Greece.

21 As Diodorus of Sicily records in his *Biblioteca Historica* (books I and V), the Egyptians quickly made use of this revolutionary mechanical implement for the irrigation of fields that lay beyond the reach of the Nile floods. The spiral pump (also called Archimedes' screw pump) is composed of a tube with internal wooden screws, whose lower extremity lies in the water. Rotating the screw with a handle causes substantial masses of water to be forced upwards.

22 The terms 'Greek Sicily' and 'Sicelian Greeks' will be use in this chapter refer to the period between the 5th and 6th centuries BC, when the island was heavily influenced by the political and cultural life of ancient Greece.

held to be a 'Commons': it was part of the *demosìa*, that is to say a right for all citizens (Tölle-Kastenbein 1990, Altamore 2008).

Indeed every large Hellenic *polis*[23] was equipped with aqueducts for the distribution of public water. Once the water reached the city centre, it was distributed to the various fountains (*krene*) which were accessible to all citizens. From these fountains water flowed abundantly, spurting from multiple taps or from elegant mouths or openings, often in the shape of a lion's head which recalls the name of the 'guardian of the water fountains' (*krenophylax*) given by the Greeks to the king of beasts (Tölle-Kastenbein 1990, Altamore 2008).

Water was considered sacred in Greece in many ways. Water that had been 'sent from Zeus' ('*ek Diòs*') was seen as having multiple beneficial properties, not least that of fertilising the earth through rain, thus ensuring health and prosperity. Famous oracles, such as the Apollo of Delphi, were closely connected with water sources. As Chamoux (1963) has argued, the Greeks themselves, on seeing certain manifestations of water, were struck with *thambos*, a word which denotes awe at the mystery and sacredness of life, a phenomenon described by classical poets. Water was simply indispensable for all ritual activity in the ancient Greek world, since it allowed for the vital act of purification. Initiation into many cults was preceded by catharsis, in which both body and spirit were cleansed. On a more general note, all purification rituals which preceded sacrifice had to be performed either with running or sea water (Detienne and Vernant 1979).

In Greek Sicily, water sources and rivers were greatly revered. As Sophie Collin Bouffier (2003) has pointed out, it is not by chance that the point of anchorage of Greeks in Sicily, the basis on which they founded their existence, is always the waterway or source around which they erected their colonial cities. In order to better understand the intimate and sacred relationship of the Sicilian Greeks with water, in this chapter I will analyse the theme of personification or the apotheosis of springs and rivers – a custom that seems to have been widespread in the entire Hellenic area.

It is well known that in ancient Greece the flow of a river was envisaged as having the earth-shattering strength of the bull, and as such, a divine force. The archetype is the personification of the River Acheloos as a bull with a human face. Also, in the Iliad, the River Scamandros pursues Achilles 'bellowing like a bull'.[24] According to the historian Aelianus, the inhabitants of Syracuse depicted the river Anapo, which flows through the city, as a man of mature age. The local springs of Ciane and Arethusa, in contrast, are depicted as female nymphs. The inhabitants of Agrigento 'make sacrifices to the river named after the city, depicted as a beautiful child': to this end a gold and ivory statue of a child was consecrated at the sanctuary of Delphi, upon which the name of the river was inscribed.[25]

---

23 'City.'

24 In the Iliad again, sacrifice of bulls to rivers is mentioned in XXI.37 and XXI. 130–2.

25 Aelianus, *Varia Historia*, II.33.

The deification of springs is also a well-known theme in some of the most famous myths related to Greek Sicily. The myth of Alpheus and Arethusa made Syracuse famous in much of the antique world for the special relationship it had with the spring of Arethusa, whose fresh waters spring abundantly just a few metres from the Ionian Sea. References to the source of Arethusa, which still enchants visitors to the island of Ortigia,[26] in the heart of the Mediterranean, are numerous in written records. According to the myth, Alpheus, the river of Olympia (a city consecrated to Zeus) and the personification of a river god, fell in love with Arethusa, nymph of the goddess Artemis, and tried to woo her at all costs. Arethusa, in her attempt to flee his embrace, invoked the protection of Artemis, who turned her into a fountain which plummeted into the Ionian sea, only to mysteriously burst forth again at the shores of Syracuse. But Alpheus would not be beaten, and he too sank into the Ionian, contriving to mingle his waters with those of the beautiful nymph.

In spite of the different versions of the Arethusa myth that have been passed down to us, its political significance would seem irrefutable: to affirm that the water fountain of Syracuse has a (symbolic) origin at Olympia, in the Peloponnese, connects it directly with the Greek world. According to Pindar and Timeo, the waters of Arethusa actually became murky and tinged with red following the sacrifice of bulls at Olympia, in the Greek homeland (Collin Bouffier 2003). As Malkin (1987) has noted, this myth would therefore seem to confirm the official status of Syracuse as a Hellenic *polis* – a special title conferred on the city by no lesser god than Zeus himself.

As Collin Bouffier (2003) argues, a well-constructed body of myths centred on rivers and water sources was necessary to the colonies in order to justify a certain political stance, both towards the indigenous colonized population and towards the Greeks themselves. By adopting the name of the rivers near which they originated, the Greek colonies of Sicily were indeed claiming a sort of fluvial incarnation to 'establish their political, cultural and religious unity.' To this end, they chose the image of the river or source to identify the *polis* on their official emblems and icons, as we shall see when we consider coins and *ex voto* made on the island.

Another fascinating myth that links Syracuse to its own water sources is that of Ciane, which Ovid narrates with special attention to detail.[27] The Syracuse water nymph, Ciane, attempted to impede the abduction of Kore, the beautiful daughter of Demeter, by Hades, God of the Underworld. In response, Hades struck Kore angrily and split open the earth creating a precipice, near Syracuse, into which he and his divine bride plunged. As punishment for her proud opposition, Ciane was turned into a spring. Diodorus of Sicily too confirms that it is precisely where 'Hades and his prisoner (Kore) plummet into the underworld' that the spring of Ciane, which he calls 'the obscure',[28] bursts forth. The different versions of the

---

26 Ortigia is the most ancient part of Syracuse.

27 Ovid, *Metamorphosis*, V.409–437.

28 Diodorus of Sicily (79-25 BC), *Biblioteca Historica*, books IV (23–24) and V (3–4).

myth confirm both the connection between sacred water and the chthonic world, and the major ritual implications for the civilian body of the Greek colony. Indeed Diodorus' version informs us that every year, in the vicinity of Ciane, 'the people of Syracuse celebrate a public *panegiris* during which small sacrificial victims are offered, while the *polis* sacrifices its best bulls to the water of Ciane'.[29] The sacrifice of bulls, which were then immersed in the Ciane swamp, was the ritual act that symbolized and re-enacted the descent (*katagoge*) of Kore to the Underworld, a descent which made way for the rebirth of the cycle of nature. From this point of view, as Faranda (2003) argues, a connection between the 'sacred' waters of Ciane, productive cycles and fertility is clearly apparent. The myth of the abduction of Kore reflects the cyclical nature of agriculture, characterized by a seasonal rhythm in which light and dark, the life and death of nature, alternate in the *kosmos* of productive cycles.

But the Sicilian historian gives us a further interesting piece of information. According to him, all nymphs of the island water sources were originally 'indigenous forces' who were placed in the service of the gods of the Greek pantheon after a journey Hercules made to Sicily. It was indeed this semi-divine hero who 'taught the indigenous population to make sacrifices to Kore each year ... magnificently, facing the spring of Ciane' (Collin Bouffier 2003).

As Bayet (1923) has noted, Hercules, Greek hero and simultaneously *dieu des eaux infernales*,[30] is very often presented as the founder of cults nearby springs, lakes and swamps, this unfailingly under the patronage of Kore-Persephone. Therefore the myth of Ciane would further confirm the important religious orientation of the island, famously consecrated to Demeter and Kore-Persephone. The complementary goddesses would have combined the dual underworld-agrarian aspect related to the cycle of life and rebirth[31] in such a way as to guarantee abundant harvests and the overall prosperity of Greek-Sicilian colonies.

In terms of iconography, it is interesting to observe how the depiction of rivers and springs is a characteristic feature of the early Greek-Sicilian coinage. According to Jenkis (1970), this may be divided into two principal types which derive from the Homeric model. In the first type, the river is depicted as a bull, which may also be a bull with a human face. In the second type, the river rather appears as a human entity, with little horns or with bull's ears.

The first Greek-Sicilian colony to adopt this coin iconography was Gela, where a bull with a human face begins to appear from the beginning of the 5th century BC. Catania and Selinunte were soon to follow suit. The humanization of rivers (second type) predominates only towards the end of the 5th and the beginning of

29 Diodorus of Sicily, *Biblioteca Historica*, IV.23.

30 God of infernal water.

31 Demeter, the divine mother of Kore (also called Proserpina or Persephone) is the goddess of the land, protector of agriculture and cultivations. Her cult worship is widespread principally where abundant wheat grows, particularly in Sicily and the Eleusian plains of Greece.

the 4th century BC. We see the head of the river personified as the head of a youth (with bull horns and/or ears) both in the aforementioned colonies, and at Agrigento and Naxos. From the end of the 5th century BC, Syracuse adopts a female head, surrounded by fish and dolphins, as the personification of the fountain of Arethusa. As Collin Bouffier (2003) has argued, it is no doubt significant that all the main Greek colonies which had this special relationship with their rivers and water sources also applied it to their coins, as an emblem of the city itself.

Archaeological excavations also confirm the sacred nature of rivers, springs and fountains in Greek Sicily. In many temples and sanctuaries erected in their vicinity, a great number of votive deposits have come to light, with thousands of artefacts including water vases (*hydrie*) and terracotta *ex voto*. Statuettes of women with children in their arms have been found at the sanctuaries of Fontana Calda and Palma di Montechiaro along with offerings and sacrificial piglets. Some of the votive inscriptions have actually allowed for the identification of the divinity to whom the cults were dedicated. At Gela, the cult of Bitalemi reports dedications to Demeter Tesmophoria, a goddess skilled in guaranteeing fertility as well as the continuity of the civil corps of the colonial *polis*.[32] In the sulphurous caves of Mount Kronio, a female divinity seated on a throne has been found, akin to Demeter or Kore. At Palma di Montechiaro and Fontana Calda, numerous female terracotta statuettes have been found which are evidence of the Sicilian cult of Artemis, depicted with a mantle (*himation*) and accompanied by a dog, deer or panther. The consistency of such finds in the Sicilian votive deposits all lead to an interpretation of the various cults associated with water in terms of their strict connection to fertility and the chthonic world (Coarelli and Torelli 2000, Collin Bouffier 2003).

From what has been analysed to date through literary, iconographic and archaeological sources, certain conclusions may be reached about the religious conceptions of water that are common to Greek Sicily. First, as the most ancient coins suggest, rivers and fountains may be seen as emblems of the colonial city itself and in some cases symbolize the intimate connection of the colony with the Greek homeland. Secondly, archaeological research has demonstrated that in Greek Sicily the water of the gods is conceived of as an essentially underground source, closely linked to fertility and agricultural productive cycles. Lastly, if it is true that rivers are often deified, it is also evident that their power is considered inferior to that of the Olympian Gods. The predominance of divinities such as Kore, Demeter, Artemis or Apollo in the temples associated with fresh water would confirm indeed that rivers and springs are always personified as minor gods in the Greek Pantheon. As Collin Bouffier (2003) has argued, the manifold Greek Sicilian water cults were actually dedicated to the more important Greek divinities,

32 The Tesmophorie were rites celebrated in the entire Hellenic world during the period of ploughing and consisted of several different ceremonies. The Tesmophorie also testify the processions of women who brought *ex voto* in the form of womb shaped pasta, symbol of fertility, to the sanctuary of Demeter-Kore.

such as Kore and Demeter, Artemis and Apollo,[33] or Zeus and Hera, rather than to local nymphs or eponymous gods. This was certainly true as long as the authority of the Greek *polis* was secure.

In the reconstruction of Greek-Sicilian water cults, however, it is important not to underestimate the diachronic dimension: the persistence of certain rituals over more than a thousand years may have implied various forms of syncretism. For example, when the authority of the official Greek pantheon began to decline, we witness the diffusion of the goddess Artemis accompanied by a piglet (as sacrifice). During the late Hellenistic period, the appearance of *paides*[34] and Maenads, are evidence of fusion with new water cultures. Also, as Brelich (1964–5) has argued, various forms of syncretism between the indigenous gods of the island[35] and the Greek pantheon (still in formation in the 5th and 6th century BC) can be found even during the period of the island's colonization. New cults, in keeping with the sacred nature of water, seem therefore to have supplanted old cults over a period of almost a thousand years, inspired, not least, by the shimmering instability of the precious, life-giving liquid.

Rome, Regina Aquarum

In the ancient Mediterranean world, the union of water and urbanisation found its maximum expression in the fountains, aqueducts, spas and sewerage of Rome. The colossal system of aqueducts and sewers constructed in the city are emblematic of a vast technological and creative expertise, one that produced some of the most outstanding and long-lived hydraulic works in the ancient world.

Between 312 and 226 BC, 11 large aqueducts were built in Rome alone, veritable marvels or *mirabilia*, admired over the centuries by both travellers and scientists. In the 2nd century AD, the Greek geographer Strabo records that 'the quantity of water conducted into the city is so great that in the underground canals actual rivers run'.[36] It has been calculated that the entire network of the principal ducts extended over an area of more than 500 km, and that at its optimum, the capacity of the network to provide drinking water could be in excess of six cubic metres a second. According to Trevor Hodge (1992), the Roman aqueducts of the 1st century AD provided the city with more water than is estimated for the city of New York's water supply until 1985. When the population of imperial Rome reached more than one million, this system assured each inhabitant almost 500 litres of drinking water a day, a quantity that is not inferior to the average *pro*

33 Due to the intimate relationship of the divine brothers Artemis and Apollo with purity, these gods are connected to all running water sources. Their temples are very often erected near springs, fountains, rivers or the sea (Tölle-Kastenbein 1990).

34 Female children.

35 Greek records do not identify these divinities more precisely.

36 Strabo, *Geografia*, V. 3–8.

capite consumption in today's major European and United States. Still today, Rome carries all the reminders of this glorious past: there is no square without a fountain, a monument to the source of life itself.

Rome in its heyday had 1,352 water basins and fountains, 856 public baths, 15 *ninfei* (fountains), 11 aqueducts, 11 large spa complexes, and five *naumachia* (naval combat) stadiums. The extent of the two longest aqueducts, the *Aqua Marcia* and the *Anius Novus* was 91 km and 87 km respectively.

The highest official in charge of this vast hydraulic complex was the *curator aquarum*, who answered directly to the emperor and whose job was to watch over the quantity and the quality of the water supplied. Below the *curator*, about 700 people were employed in the administration of this large water network. Such a system has led many scholars to consider Rome the best-served city in the ancient world. With the Romans, the history of water and that of urbanization merges definitively (de Klein 2001, Altamore 2008).

Perhaps the most complete picture of Roman aqueducts that has been passed down to us is that given by Frontinus, who was appointed as *curator* in 97 AD. His *De aquaeductu urbis Romae* is considered the most important scientific and documentary work on the hydraulic architecture of the ancient Rome (Rodgers, 2004). The first systematic description of hydraulic engineering dates, however, to Vitruvius (1st century BC). In his famous account of the state of knowledge during the early Augustan period, *De architectura*, Vitruvius gives us some very detailed descriptions as to methods of water supply (water retrieval, purification and distribution), as well as its possible use for hydraulic machinery (mills, large water wheels and other techniques for the lifting of water)[37] (Repellini 1989).

Water in Rome, as in Athens, was considered a *res publica*:[38] its availability was guaranteed to all citizens free of charge. For the masses, water was available from the many fountains of the public streets and squares, thanks to the ingenious methods of distribution which assured a regular and effective supply to all. Public law upheld the principle whereby no citizen could be denied this precious resource. Roman aqueducts were public (property of the government). Indeed, Roman law provides an important normative contribution to the founding of a 'water culture' aimed at its administration in the interests of all citizens (Altamore 2008).

Those who benefited from Rome's water power and ensued improvement in general hygiene were not only the Romans but also those populations conquered by the ever-expanding empire. There are important remains of Roman aqueducts in many parts of Europe and the Mediterranean, including the spectacular Pont du Gard (50 AD) composed of three levels of elegant yet powerful superimposed arches, 48 metres high and 490 metres long. It is the highest of the Roman bridge-aqueducts to survive to the present day, and was part of the 50 km long aqueduct of Nîmes (France). Among the other outstanding examples of these large hydraulic infrastructures are the 132 km long aqueduct at Carthage (Tunisia) and

37 Vitruvius, *De architectura,* books VIII and X.

38 'Public property.'

the aqueducts at Cologne (Germany) and Lyon (France) whose respective length was almost 80 km.

Roman aqueducts worked on the principle of gravity, making use of a small but regular slope between the collection source and the destination. Roman engineer-architects became truly adept in these constructions. At that time, they used sophisticated equipment like *choròbates*, to measure the earth's inclination, and dioptres, to make underground canals and tunnels.

A large number of monumental dams, reservoirs, and aqueducts were also constructed in Spain, North Africa and in the Near East (Libya, Syria, Jordan, Lebanon, Israel and Turkey). In the 6th century AD, under the Emperor Justinian, the Romans constructed a large underground reservoir of drinking water in the heart of Constantinople supported by a thick forest of stylish columns. This famous basilica-cistern, which may still be visited today, exemplifies the advanced techniques and expertise of a Roman hydraulic engineering tradition that would spread throughout the Holy Roman Empire.

In the Roman Empire we also find the systematic application of a very efficient system of water distribution called the water tower (also known as 'castle' or *castellum*). The water tower allowed for the establishment of a very precise hierarchy amongst all users, thanks to a system of three concentric basins placed at various heights (the so-called 'waterfall method'). When water was scarce and not all of its uses could be guaranteed, the supply from the lowest basin was reduced, because the water that flowed out of the top basins through spill-over was diminished. The *castellum* thus automatically fixed the priority of supply first and foremost to the public fountains and pools (connected to the upper basin); second to the spas and theatres, supplied by the middle basin; lastly, the private houses, connected to the lower basin (Repellini 1989, Altamore 2008).

Not only are the construction techniques used for aqueducts and water distribution extremely well documented, but also the filtration and purifying systems for drinking water. The water collected from springs, lakes, rivers and drainage tunnels passed through a cistern whose function was to create sediments of sand, gravel and mud: the *piscina limaria*. The water was additionally purified by passing through a bed of quartz sand (a system which is still in use today in some modern plants).

Seeking out the best source of drinking water was a major preoccupation in the Roman world, so much so that Vitruvius states: 'it is vital to choose well those water sources that are healthy and life giving.' In his *De architectura*, he considers the question of the best sources, giving very detailed descriptions as to their location, and also as to water types and tastes.[39] The *Naturalis Historia* by Pliny is also full of references to natural springs, hydromancy, and the use of water in medicine. According to the historian, 'waters are akin to the territory through which they flow.' The properties of water, including those of air filtration,

39 Vitruvius, *De architectura*, book VIII.

and changes in its chemical composition according to characteristics of the soil through which it flows were well known to the Romans (Repellini 1989).

At the apex of the Empire, all principal Roman cities had a good water supply system as well as an efficient sewage system. This parallel system for the collection of sewage and used water assured that the urban areas had maximum hygiene, an unprecedented achievement at the time.

Rome's sewage system, based on the *cloaca maxima*, seems to have been of Etruscan origin. Its construction is presumed to date from the time of the Etruscan king, Tarquinius the Haughty (Vegetti and Manuli 1989). Spas too were an important part of this amazing system of public hygiene, since their drainage waters were in turn used for the sewers, thus guaranteeing the cleansing of public toilets, which flowed into appropriate sewage collectors. Scarborough (1969) maintains that sanitary engineering was one of the triumphs of Roman civilization, in that it created a system of hygiene unsurpassed in the Mediterranean and Europe until the 18th century.

With the development of spas, water became the object of highly evolved therapeutic practices. Where the health-giving liquid burst forth from the earth, temples to Asclepio (the Greek Aesclepius) and other gods were erected. Hydrotherapy, consisting of hot and cold baths to remedy various ailments, was practised near the springs. Thus, alongside temples dedicated to worship, buildings were erected to accommodate patients. Indeed, in Rome, these spas grew into small urban complexes, complete with gymnasiums, libraries, shops, eating and meeting places. In some cases, the prefix *aquae* was attached to the names of some cities, both in Italy and in Europe[40] (Altamore 2008).

As many historians have pointed out, the manifold capacities and knowledge of water were fundamental for gaining consensus during the time of the Roman Empire. As part of the well documented capacity of the Romans to satisfy the basic needs of the conquered populations, the 'hydraulic lever' became an important instrument in imperial politics (Vegetti and Manuli 1989).

As for the religious aspects of water in Roman life, like the Etruscans and the Greeks, the Romans associated water strongly with the sacred. Conceived as 'bridges between two worlds', as Guzzo (2003) argues, Roman springs had manifest creative and transformational properties: they were sites of oracles and miracles, and their waters were associated with rejuvenation and fertility.

In Rome, both the *Fontinalia* holiday, its temple (probably set at *Porta Fontinalis*), and the god associated with it (Fons or Fontus), are clear examples that water was held sacred. Frontinus records that until 312 BC (year in which Rome's first aqueduct, the *Aqua Appia*, was completed), the Romans took their water from local wells and water sources, which were held to be sacred and capable

40 The German Aquisgrana (from *Aquae Grani*), French Aix-en-Provençe (from *Aquae Sextiae*) and English Bath (from *Aquae Sulis*) are clear examples of how the Roman term was retained in the name of the city itself.

of restoring health to the sick.[41] Also, ancient Roman religion speaks of the water nymphs Camene (known also as the Muses in later periods), who were honoured at *Porta Capena*, their principal place of worship, with libations of water and milk (Tölle-Kastenbein 1990).

Water was fundamental not only in libation and in all other daily purification rites (*lustratio*) but a central element of the Empire's main ceremonies. On returning home, the victorious Roman armies were welcomed with a shower of sulphurous water. The name of the goddess Giuturna (perhaps of Etruscan origin) was given to the fountain of the Roman Forum whose water was used for the principal sacrificial ritual of the state. The Greek cult of Zeus Liceus ('bringer of rain') corresponds to the Roman *Aquaelicium*, originally a magical practice to invoke rain that was later associated with Giove Capitolino ('dispenser of rain'). It is notable that the title of *pontifex*, which was given to those who were assigned to create bridges and passages from one side of a river to the other, has maintained its sacred connotations to the present day (Tölle-Kastenbein 1990).

It also seems certain that the kings of Rome were themselves linked to some form of personal water cult. There is a legend that Numa Pompilius, who reigned from 715 to 674 BC, had a nymph called Aegeria as a counsellor. The French geographer Elisée Reclus drew on this legend for his *Histoire d'un Ruisseau* (1869), in order to re-state the intimate connection between water and culture at the origin of all civilisations. Reclus writes that Numa Pompilius,

> used to wander alone into the depths of woods and [...] when he came near to the sacred cave, the purity of the waterfall, with its dress fringed in foam, and the moving veil of iridescent vapour, in his eyes seemed to take on the form of a beautiful woman. He, poor mortal, spoke to her and the nymph replied with her crystalline voice. [...] Thus he learned what wisdom was. No old, white-bearded man would ever speak words similar to those of the nymph, immortal and ever young. [...] The names Numa Pompilius and Aegeria summarize a whole period of the Roman culture. Humankind, at the origin of every civilisation, derives its laws and customs from the nymphs, or better, from nature's springs, mountains, and forests.

The diversity of rituals practised in such a vast empire as the Roman cannot be further investigated here. It is also true to say that no one single religion was ever valid for subjects throughout the empire, and the only religious principle deemed acceptable by the Romans was that of polytheism and multi-religious practice (Scheid 1989). With the principle of *cuius regio eius religio*,[42] Rome not only allowed for the continuity of local cults and religions in each corner of the empire, but also assimilated a considerable number of rituals and ceremonies associated with water into its own religious culture. Thus, the sacred connotations of water

41 Frontinus, *De aqueducti,* I.4–5.

42 'To each region, its religion.'

existent in the empire should be viewed in terms of their infinite variety, not least in the light of the extraordinary open-mindedness shown towards all cultures of the wider Mediterranean basin.

Water and Christianity: Continuity, Transformation, Survival

With the decline of the Empire, Rome's vast water system began to fall into disuse. When the emperors abandoned the city, the aqueducts were no longer looked after and the city's water dispersed. Fountains dried up, and aqueducts and spas gradually stopped functioning. With the advent of Christianity, the pagan divinities associated with those places where hydrotherapy was practised were progressively censured because of the suspicious climate of promiscuity which had become associated with these complexes. In addition, the cults and pre-existing divinities linked to many water sources were often demonized. There are clear traces of these ancient divinities in Italian folklore, including legendary figures of nymphs, *paides* and *anguane* which Christianisation proceeded to cast as diabolical.

At its very beginning, Christianity in turn became the source of a new sacred conception of water linked to the rite of baptism. The expression 'living water' summarizes the spiritual character of baptism. Christ himself provided one of the most exemplary pieces of evidence for the sacredness of water, when he said 'whosoever drinketh of the water that I shall give him shall never thirst; but the water that I shall give him shall be in him a well of water springing up into everlasting life' (John, IV.10–14). In this sense, water is often held to be the symbol of the Holy Spirit (Ravasi 2005).

An extremely rich symbolism finds expression in the ritual of baptism. In the early years of the Church, adults were baptized by total immersion in water. This immersion and subsequent emergence symbolized spiritual death and rebirth: the immersion was conceived as a 'dying with Christ', a descent to hell which was also conditional for rebirth; the emergence was a sign of liberation from death – it implied the 'creation of a new human being.' In this sense, baptism is the pillar of Christian liturgy. As Caiazzo (2003) argues, any other use of water in the Christian religion is derived from and takes its full meaning from the rite of baptism.

The advent of Christianity also saw the substitution of pagan temples linked to water sources with new buildings dedicated to the Christian faith. In Italy there are a great number of churches and monasteries erected on sites where pagan cults connected to the various manifestations of water once existed. According to Sébillot (1983), of all the archaic religious rites, the cult of water is the one that has best survived to the present day. Anthropologists, folklorists and scholars of popular traditions have often stressed the aspect of 'survival' of water cults, even if recognizing the very profound changes at a social, cultural and economic level that have occurred over the centuries. In many cases, as Hobsbawm and Ranger (1983) have pointed out, this emphasis on formal resemblance and continuity has overshadowed the continual re-invention of traditions that often accompanied so-called continuity. Thus, it is important to verify what might seem to be continuity with rigour, case by case, and not simply to generalize.

With this methodological precaution, it is possible to consider some representative examples of sites where water has been held sacred over an uninterrupted period of time, from prehistory through the Medieval and the Renaissance periods to the present day. This is testified by the continuity among Etruscan-Roman remains, Romanesque churches and other historical Christian monuments.

In Valdambra and Valtiberina,[43] many votive objects found near wellsprings testify to their use over a period spanning more than three thousand years. Near the San Leolino *fonte lattaia*,[44] which was considered until very recently, according to oral testimony, to have miraculous benefits for breastfeeding mothers, a prehistoric artefact was found with very pronounced buttocks and breasts. According to Dini (1995), this is most likely a votive offering to a female goddess or to the water source itself, in accordance with ancient rituals of devotion. At Monterchi, in the area where the church of Santa Maria di Momentana stands, there is evidence of a water cult associated with health which has lasted for a long period of time. This area is characterized by several prehistoric and Etruscan-Roman settlements, as well as by springs that were renowned for their therapeutic qualities especially pertaining to sterility and difficulties in pregnancy. Votive artefacts belonging to the Etruscan-Roman period have been found near the church (including terracotta breasts and wombs). Perhaps it is not by chance that Piero della Francesca painted here, in the Chapel of Santa Maria di Momentana, one of his most mysterious frescoes, the Monterchi Madonna of Childbirth (1455). This work depicts Mary in a way that is very unusual for Italian Renaissance painting, that is, in a state of full pregnancy. Art historians have not hesitated to claim continuity between the Christian Virgin and Etruscan-Roman divinities like Uni and Giunone Lucina, who protected pregnant women and childbirth (Dini 1995, Cretella 2005). Certainly this masterpiece seems to confirm the pre-existence of local water cults.

There are several other well known maternity sanctuaries in Tuscany, among which are Mount San Savino and Lucignano in Valdichiana. Here too, there is enough archaeological evidence to assert that ancient water and health cults survived, despite various and obvious transformations over the centuries, to the present day (Dini 1995). Calabria, the Italian region situated in the southernmost reach of the peninsula, also merits special attention for the evident transition of a Magna Graecia water culture from ancient times to the modern era.

43 Etruscan areas of Tuscany.

44 Milky spring.

The Thirst of the Dead. Dangerous and Prodigious Waters in the Folklore of Calabria

Saints capable of offsetting incessant and ruinous rain, or of bringing prodigious and miraculous water after long periods of drought, are deeply rooted in the traditions of Calabria. In this area, historians and folklorists have gathered a vast repertory of beliefs, legends and rituals connected to patron saints who were invoked to tame the wild force of rivers and to control extremes of rainfall, both too much or too little. Here I shall look at some of the most significant evidence gathered on this subject. A profound link between water and the afterlife or the world of the dead is also found in relation to customs and practices surrounding the myth of the 'great thirst of the dead', which I will consider later.

In several parts of Calabria, in order to stop incessant rain, it was customary to throw two pinches of salt into the fire and another seven out of the window in the direction of the clouds, while reciting: 'water go / Saint John sends you away / into the Lord's mind.' Invocations of the local patron saint to stop the rain were very widespread. At places like Monasterace and San Giovanni Theriste, invocations of this type during thunderstorms and flooding are still in living memory. When prayers and imprecations were not enough, there is evidence of the very curious custom of parading the saints and punishing them, a custom related to the more atavistic *do ut des*. At Nicotera, for example, the statue of San Giuseppe was taken to the sea shore, heaped with insults, and left there until the rain stopped (Angarano 1973, Teti 2003).

Other examples of this 'disrespect', which in some cases continued until the 1950s, confirm that similar customs were used in times of prolonged drought. The statue of the saint was stripped of its regalia, tied with ropes and left naked in the middle of the church until rain came and thirst was quenched. Both at Riace and Squillace, the statues of saints were given a punitive bath in the sea, and if the rain still failed to appear, insults were hurled at them. Similar customs were practised at Paradisoni, Rossano and Malvito, where San Pietro, San Nilo and San Michele were venerated respectively (Angarano 1973, Teti 2003). This punishment of bathing saints has also been documented in other countries such as France (Sébillot 1983).

The folklore of Calabria also has many echoes of the theme of the miraculous wellspring. Miracles associated with water often occurred after the death of Italic-Grecian saints. According to a common belief, holy water burst forth directly from the tomb, the relics, or the hermitage in which the saint lived. Legends and traditions also mention miracles associated with the Holy Virgin, who miraculously made water appear in areas of aridity, or who appeared herself near springs and fountains. At Bombile, the Madonna of a cave near the Miraculous Fountain is venerated; at Spilinga there is evidence of a cult of the Madonna of the Fountain, so named because of her apparitions near a local spring.

Rites of exorcism are also linked to the waters of Calabria, as in the case of Africo. In the ritual that is linked to San Bruno, water itself is a vital mediator in the healing of those possessed. Tradition has it that San Bruno, near the present day monastery of Serra San Bruno, macerated his body in a small lake, and through his penance sanctified these waters, which then became the central element of the rite. The custom of exorcising demons takes place on the Monday of Pentecost and consists of plunging the possessed in the lake to free and purify them (Ceravolo 2003).

Many interesting customs are known to have existed in Calabria related to the theme of 'the thirst of the dead', a myth that occurs both in Mediterranean and Euro-Asiatic folklore. In the Mediterranean, the myth surrounding the insatiable thirst of the dying and the dead dates at least to the Mycenaean world, where the term 'thirsty' (*dìpsioi*) was used as a synonym for the dead. Also, in Mycenaean culture it was customary to bury the dead in a bath, so as to quench their *psyche* (spirit) (Pugliese Carratelli 2001).

As for the evidence from Calabria, the 'thirst of the dead' theme revolves not only around the departure for the afterlife, but also the temporary coming back to life of the dead. The inhabitants of Celico, Trebisacce and Acri, as reported by Dorsa (1884), used to leave a piece of stale bread beside the deceased as well as a mug of water to quench his thirst, making sure that the door of the room was firmly closed. In fact, if observed, the deceased would not eat the bread or drink the water, and his soul would wander disturbingly throughout the house. In this case, only special rites of atonement could offset the possible dangers involved and lead the deceased to the extreme place of unction. During funeral ceremonies, this belief is also demonstrated in many places, as exemplified by the custom of pausing beside fountains on the way to the cemetery to allow the soul of the deceased to drink for the last time before finally being laid to rest (Teti 2003).

Another widespread custom was that of leaving food and water out for the dead on the nights between Christmas and New Year. Their return thus occurred in an 'organized' way, limited to certain days and occasions. Adhering to fixed rituals, this return from the grave was regulated by providing water, used to prevent unexpected and unwanted apparitions. In this sense, water delineates the separation between the living and the dead; it both accompanies the deceased on his final journey and acts as a medium through which he can return.

The uncontrolled, if not uncontrollable presence of dead souls, moreover, was believed to exist in the vicinity of fountains, springs, and rivers. According to the topography of folklore, as Lombardi Satriani and Meligrana (1982) have pointed out, fountains are 'places where restless residing spirits can attach themselves to passers by.' In order to avoid meeting the spirits of the dead, one had to stay away from these water sources, especially at night, when the moon was full, or at other 'critical' times of the year. Water, therefore, becomes an irreplaceable medium through which the 'terrible nostalgia for life' – in a final analysis an unquenchable thirst – is appeased (Sébillot 1983, Teti 2003).

From the cases we have discussed, a precise cognitive map of water symbolism may be drawn: a map which is characterized by the very typical and inherent ambivalence of the sacred. Indeed water takes on a myriad of opposing meanings and, depending on the circumstances, can be both pure and impure, good and evil, capable of exorcising the dead and yet constituting the ideal habitat for infernal spirits (Ceravolo 2003). In the folklore of Calabria, and of the Mediterranean more generally, water has precisely this characteristic: it may take on diametrically opposed meanings according to the times and contexts in which it is called on to appear.

Modern Myths: From Technocratic Paradigms to Bottled Water

Our examination so far amply confirms the existence of several unique water cultures in the ancient Italic world. The diversity of these cultures is held together by a single common denominator, and that is the way in which water was conceived. First and foremost, in all these cultures water was conceived as a precious 'good', very often associated with the sacred, and simultaneously also as a resource that is not endless and has to be guarded with great care. How many of these conceptions so imbued with respect for water and its preservation still survive in modern Italy?

In this final section, I will consider how new behaviours and ideologies related to this vital element have developed over the last few decades. Indeed, the most common attitudes to water today seem based on a 'culture' of waste, pollution and disrespect that is precisely the opposite of what existed in the past. From this perspective, I will examine some of the most evident contradictions that characterize today's usage and perception of water, from the misunderstanding of critical consequences that the present 'development model' has produced in all surface and underground water, to the common conceptions about bottled water – of which Italy is the biggest consumer in the world.

In Italy, during the first half of the 20th century, water started to be available in all homes in virtually unlimited quantities. Simultaneously, however, the perception of the fundamental importance of water vanished from people's minds. How many people today consider the intrinsic value of running tap water? Practically no-one, because 'water, like all things considered only as a commodity, has become alienated from its own history and made invisible.' In modern Italy, as Franzin (2005) has pointed out, water has lost not only its aura of sacredness and respect, but its history too.

Until the first half of the last century every water source, fountain or stream had a name, a history, a legend associated with it that spoke of the mentality, perceptions and representations of the local communities. These water sources were unfailing reference points for all kinds of activity: special places which also become points for gatherings, exchange and other kinds of social activities (Teti 2003).

This picture has radically changed in the present day. Piped, hidden and polluted water has progressively taken the place of free and flowing water. Water has even slowly disappeared from the landscapes and regions in which we live, only to suddenly reappear or disappear in case of emergencies like flooding and drought. But apart from the undeniable advantages of piped water, what negative consequences has this development model had on the environment, the landscape, and indirectly on human health?

The main sources of water pollution in present day Italy are urban and industrial waste as well as agricultural waste. During the last 50 years, water containing all sorts of organic material, fertilizers, pesticides, industrial waste, detergents, petrol, metals, radioactive waste, and debris from roads and construction sites has radically impaired the ecological equilibrium of many aquatic ecosystems. Nowadays chemical and industrial waste still continues to be dumped into rivers and underground water, often without any purification treatment. All of these substances of course create havoc with human health. Only today, after decades of large scale pollution encouraged by far too permissive laws, do things seem to be easing slightly (Bevilacqua 1996, Tamino et al. 2002).

In Italy the seriousness of the negative and cumulative impact on aquatic ecosystems has been wholly underestimated. As is well known, these ecosystems are the primary unit of the natural water cycle. They alone – if properly functioning – can guarantee the natural regeneration of the quality of free-flowing water, and this totally free of charge. Nevertheless, while billions of euros have been invested in recent decades to strengthen the artificial water network (reservoirs, irrigation canals, water pipelines and sewage systems), transforming the natural hydrological dynamics of most river basins, the natural ecosystems have been continuously and systematically exploited and forgotten (Ghetti 2008).

Aside from pollution, we must also add the continual modification of the landscape due to the processes of economic development and urbanisation, which have heavily compromised the natural flow of all river basins. The construction of dams to create artificial basins and hydro-electrical plants, the building up of river banks to contain high water, and the exploitation of rivers for agriculture, urban and industrial installations, have all too often tampered with the natural processes of water flow, and the consequences have been disastrous for the environment and the natural hydrological dynamics of water bodies. It is precisely these human interventions that have determined the modern landscape of Mediterranean Italy (Turri 1998, Franzin 2005, Vallerani and Varotto 2005).

In addition, we must not forget the abuse of all surface and underground water that has occurred in agriculture, including the exploitation of non-renewable underground water for private use by illegal drilling, a practice abetted by complaisant politics (Ruf and Valony 2007). If we add to this the progressively more intensive construction of cement jungles and the ensuing loss of biodiversity, the balance of the last 50 years of 'progress' in Italy borders on nothing less than ecological disaster. Pollution, exploitation and the trivialisation of natural ecosystems are the results of a 'development at all costs' model which needs to be thoroughly reconsidered today. Indeed such a technocratic paradigm has not taken into due consideration the negative impact on the environment in the medium and long term, in particular the cost of restoring the fragile equilibrium of aquatic ecosystems that have been deeply modified or even destroyed. Who will pay this cost? Are not we dumping the burden too heavily on future generations?

The Water Framework Directive 2000/60 (WFD henceforth), the European normative body whose purpose is to manage all European water in a more sustainable way, may offer hopes for better prospects. This ambitious Directive, which is estimated to require 15 years to establish (2000: 15), places Europe at the forefront of the world for its innovative and 'revolutionary' approach, the aim of which is to bring all water to a 'Good Ecological Status.' This goal will be attained combining three different parameters of quality: biological, chemical, and morphological. By imposing this objective on all Member States, the WFD is deliberately putting the need to restore and revitalize aquatic ecosystems at the top of national political agendas (Eulisse and Armellin 2008; Ghetti 2008).

Although the WFD undoubtedly paves the way for a new water culture in Europe, it tends to do so with little attention to the countries that border the Mediterranean and to the plurality of cultures that have historically flourished here. Essentially founded on the principles set out in Dublin in 1992, the WFD rather signals an irreparable break with the local knowledge and traditions of how water has been managed by local communities over the generations. Negative effects of European water politics include, on the one hand, precariousness for the Mediterranean small farmers, and on the other an unequal distribution of water resources for agriculture. In this process the state, far from being absent, has expropriated ancient water rights to favour precisely those who over-consume. Thus, the application of such a technocratic model is not without serious social, economic and environmental contradictions (Ruf and Valony 2007).

In ideological terms, another relevant feature of present-day development models concerns the fact that the 'technology' has become the true religion of our era and, as such, has legitimized a technocratic approach which fosters the illusion of generating only positive outcomes. As Prigogine and Stengers (1984) have pointed out, today's technology may reasonably be considered a sort of 'modern miracle': in this sense, a clear connection between science and myth may also be drawn. Science and technology may be viewed not only as having ideological affinities with religion, but also as being the engine of a new 'mythical discourse' in its own right – a discourse that has obliterated ancient water cultures and rights.

A similar modern myth marks common perceptions and attitudes towards bottled water, i.e. the 'sacred' and health-giving water of our present civilization. If it is true that technology has destroyed and usurped the sacred aura water once had, modern man – as Umberto Eco (1993) has argued – inevitably tends to create new images of purity and well-being associated with water, even if they only amount to consumer goods available in shops and supermarket shelves.

In his *Mythologies*, Roland Barthes (1957) has carefully analysed the manifestations of 'myth' in modern day consumer societies. According to him, a myth becomes a myth in as much as it expresses a particular 'way of signifying', characterized not so much by what it says but rather by the *emphasis* with which it professes its message. In a series of essays published in the 1950s, the French semiologist examines the many faces of 'modern myths', from advertising and sport to newspaper reports and political discourse.

Bottled water advertisements, from this perspective, constitute a perfect example of the forms myth can assume in the contemporary world. What plays a vital role in shaping the myth of bottled water in the social collective unconscious, is the concept of 'purity'. It is no mere chance that packaged water labels show images of the most inaccessible, natural and uncontaminated places. We see snowy mountain tops and other idyllic landscapes even when we are buying groundwater extracted from alluvial plains. Indeed in the collective unconscious, the idea of non-contamination and purity have become a prerogative of bottled water. All of this has meant that traditional public fountains and tap water are viewed with increasing suspicion, while the sealed bottle – with its 'magic powers' – is more able to satisfy the fantasies of consumer society (Macbeth 2003, Altamore 2008).

The rampant though not always justifiable success of bottled water is upheld by the idea that only uncontaminated and 'pure' water from bottles can help people to lose weight and be healthier. In Italy, this success has been fuelled by a very powerful lobby, so that millions of people have started drinking bottled water exclusively, even when tap water is better and more carefully controlled than the sealed container. These misconceptions have had serious social and environmental consequences.

Italy today is the largest consumer of bottled water in the world: 194 litres per capita in 2007 (eight times the world average). It has been calculated that in the last 15 years the consumption of bottled water has risen by 76 per cent. During the same period, the advertising investment by companies in this sector has increased massively, amounting to about 390 million euro in 2003 alone (Altamore 2008, Martinelli 2008).

This dismal record is accompanied by serious consequences for the environment. In Italy each year something like 350,000 tons of refuse is produced by PET containers. The Italian consortium that retrieves plastic packaging recycles only about a third of it. PET containers may constitute a saving for the companies, but since Italian regions bear the costs of plastic disposal, the burden then falls on the environment and on public spending (i.e. on the ordinary citizen). Huge amounts are involved: in Lombardy alone about 20 to 25 million euro a year is spent on the disposal of plastic bottles (Altamore 2008, Martinelli 2008).

The social and environmental damage of the wholesale extraction of water carried out by bottling companies must also be considered. For example, in the province of Cuneo, the multinational company Nestlé (the largest selling group in Italy) pays the ridiculous price of 2,528 euro per year for a concession that allows them to extract water from an alpine area of 67 Ha. In the province of Biella, the concession granted to the same company brings 8,620 euro in public saving but this is at the price of sanitary surveillance of the same sources, which costs the province more than double the amount (17,600 euro per year). The fourth largest selling company in Italy, Ferrarelle, pays less than 1,000 euro per year to drain the water of the same name in the province of Caserta. Examples like this are almost too numerous to mention. An examination of concessions given to bottling companies reveals that 12 out of 20 Italian regions make these concessions at ridiculously low prices compared to the colossal profits the companies actually make. Moreover since water, by law, is a state or rather a citizen-granted good, the concession entails considerable financial and environmental damage for the local communities. In many Mediterranean Italian regions the almost uncontrolled draining of water has completely exhausted water sources (Altamore 2008, Martinelli 2008).

The myth that maintains that packaged water is necessarily better, purer and safer than water from the aqueducts is based on a very slim grounds. This is confirmed by the European infraction[45] Italy incurred in 2002, which stated that 'the Italian legislation authorizes the presence of polluting substances in bottled water of which there should be no trace.' Three years later, following the implementation of the European Directive 40/2003, which establishes the parameters of chemical substances in bottled water, the Italian Health Ministry was obliged to withhold the sales of 126 brands of bottled water because of the irregular substances found in them. In terms of European limits, over 200 brands (out of 290 on the Italian market) were found to have levels of substances far greater than the permitted amounts (Altamore 2008).

In spite of these facts, sales of bottled water in Italy have not dropped; if anything they have increased. The national media practically ignored the European infraction. Even the sentences given by the Antitrust[46] to bottling companies for misleading advertising are largely ignored. Curiously, the accidental contamination and irregular parameters of packaged water receive less attention in the media than do rare accidents involving public aqueducts.

It is no surprise, then, to find that two different legislations exist in Italy regulating the use of drinking water: the first, applied to tap water, has far stricter parameters than for bottled water. On the basis of these laws, bottled water can contain substances and mineral salts in such high amounts that if they were to be tested by laboratory rules guiding tap water, they would be banned. If this water can be drunk only in bottles, it is thanks to a lobby that allowed companies to obtain a 'pleasing' legislation. Given the importance of the sector for the economy (a turnover of three billion euro per year), certain types of pressure are more pronounced than others in Italy. Therefore, water that would be considered by law 'unsuitable for human consumption' if coming from a tap, may, thanks to peculiar normative expedients and subsequent massive advertising, become 'miraculously pure'.

Faced with all this information, it is legitimate to wonder what the necessity is to consume bottled water in a country which can boast of having some of the best tap water in Europe and possibly in the world. In Italy tap water, with very

---

45 European infraction procedure no. 105.852.

46 Guarantor authority for market and competition.

few exceptions, is of good quality, safer and more controlled than bottled water; furthermore, it is available directly in the home and costs little (on average 300 to 1,000 times less than bottled water).[47] In order to explain the success of bottled water, what must be examined is evidently not the content but the container – and all that this implies in today's unconscious representations. Despite the discrepancy between scientific analysis and advertising slogans about the actual 'purity' of packaged water, its myth remains strangely intact.[48]

Traditional Knowledge for a New Model of Sustainable Development

In this chapter I outlined how in Mediterranean Italy, over the last 100 years or so, water has been downgraded from a sacred symbol of life to a mere economic resource. In Italy, as in many developed countries, innumerable forms of free flowing water have been replaced nowadays by the urban and economic mechanisms of piped water. One consequence of the rise of the consumer society is that the concept of water as a Commons, so characteristic of ancient civilisations and rural communities, has been replaced by a wholly misguided one in which water is reduced to a mere commodity, ripe for exploitation.

Italy, like other Mediterranean countries, has, in just a few decades, erased the most profound meaning and legacy of those water cultures that gave rise to its civilization, only to nurture a diametrically opposed cultural vision of water based on waste and disrespect. Thus a model of 'development at all costs' has led to the systematic exploitation and pollution of all surface and underground water. The myth that modern technology is somehow perfect and infallible has encouraged the widely-held belief that water is an unlimited resource. Today we only have to turn on the tap to have all the water we could possibly want; it is sufficient to buy a powerful water pump to illegally drain ever deeper strata in order to water crops in inappropriate places, like the semi arid environments of Mediterranean Italy. Water today is no longer managed within and by local communities of farmers, but is directly controlled by external forces, that is, by small groups of technicians, technocrats and politicians who very often perpetuate a wholly inadequate and short-sighted vision of 'progress'.

In today's Italian consumer society, thanks to massive investment in advertising, new myths about the purity of water are not forged anymore by the ancestral tutelary gods of water sources; rather, they are created by the powerful

47 In Italy 1,000 litres of tap water cost on average 1 euro. The same quantity in a bottle, purchased in a supermarket, costs 300 to 1,000 euro on average.

48 Similar conclusions were reached in the US in a study carried out by the National Resources Defence Council in 1999. In analysing both tap and sealed water, it was found that different bottled waters do not respect the rigorous sanitary limits set out for tap water. In some bottles, unacceptable levels of bacteria and other health endangering components were found (Macbeth 2005).

lobby of bottling companies, which not only invent what should be consumed, but also expropriate the culture, knowledge and places associated with water. As a consequence, the common belief is that the best and purest water is that found on supermarket shelves hermetically sealed with snowy mountain labels, rather than what flows from home taps or public fountains. The dominant perception and consideration of water that exists today in Italy fosters the exploitation of unlimited quantities of water. This erroneous vision, this seemingly insoluble cultural contradiction has been fostered by the myth of a technocratic paradigm that has erased age-old wisdom together with the concept of limitation.

The data available for the entire Mediterranean basin as to the steady reduction in availability of water for civil, agricultural and industrial use is striking. As well as problems of quantity, many countries are today facing a serious qualitative water crisis which could easily become the object of escalating conflict both locally and internationally, as the peaceful sharing of the resource becomes impossible. Strategies for coping with the complexity of the situation are often contradictory and inadequate. The Mediterranean today is facing the absolute limits of resource-exploitation. What is clear is that working towards a solution is no mere technical task or an issue of improving water technologies.

The consideration of so called 'traditional' knowledge may play an important role in future strategies to be adopted. Indeed the answer to problems of water scarcity, desertification and flooding that today plague the Mediterranean may be inspired by some principles and techniques of water management that have been used for centuries by local communities. Simple and effective know-how and knowledge such as water-harvesting, terracing, and water conservation may prove to be more effective and sustainable than many solutions offered by modern technology (Eulisse 2008). Adopting this perspective, the United Nations Convention to Combat Desertification has promoted an important census of traditional knowledge on a world scale (UNCCD 2005). Within the sphere of many development projects for Third World Countries, experience has demonstrated that much of the West's sophisticated technology, when applied in different cultural contexts, is not only ineffective, but brings with it long-term negative side-effects that may outweigh any possible benefit.

Whereas modern technology operates effectively in the short-term, traditional knowledge operates over a long period, being founded on a widespread conception of water as a Commons and with the clear awareness that resources are finite. For example, whereas certain traditional farming practices are based on the collection of rain water or make sustainable use of surface groundwater, new technologies operate by digging deeper and deeper into the ground, causing the widespread problem of salination, the exploitation of non-renewable fossil water, and simultaneously the expropriation of the ancient water rights of small farmers (Shiva 2002). Today, while the entire planet is on the brink of ecological collapse, traditional knowledge can teach us how to interact with the environment and enhance its resource potential without depleting it.

What then is meant by traditional knowledge and how compatible is it with modern technology? In the literature, traditional knowledge is often referred to as including farming practices, management models, water artefacts and even ways of behaving and attitudes. A number of studies have emphasized the tangible advantages of traditional knowledge, including the versatile nature of small scale technology that is easy to use, its low cost, and its low impact on the environment. Furthermore, traditional practices make use of renewable sources that are locally available as well as of recycling processes. Since these manifold practices have been developed within local communities which have lived in a certain area over a long period, these populations have a precise awareness of the potential and limitations of their environment. Also, traditional practices which have been experimented with over the generations in a sort of trial and error approach often entail sophisticated institutional mechanisms for the resolution of internal conflicts (Reij et al. 1996, Langill 1999, Laureano 2003).

In terms of water management, traditional knowledge has shaped unique cultural landscapes worldwide, making use of low energy and low resource-use solutions so as to adapt and react to environmental variability and emergencies in flexible and multifunctional ways. In this sense, as Laureano (2003) has pointed out, traditional knowledge may be seen as 'an organic system, comprising technical know-how and environmental awareness, solidarity among community members and the ability to manage common resources.'

It is important to note that all forms of traditional knowledge are part of a wider system, which includes non material aspects, perceptions and specific aesthetic values. Within this understanding, technical features, cultural traits and even spiritual beliefs are inextricably fused together. For this reason, traditional knowledge cannot be reduced to a mere list of technical or handy solutions that can be mechanically transposed in any situation to solve specific problems. Indeed traditional practices have been conceived in specific cultural contexts as 'multifunctional and all-purpose' techniques, i.e. techniques capable of solving several problems at the same time. For example, a terracing system is not only a means with which to protect slopes from erosion, but it is also a way to collect water and reconstruct the land. Moreover, it also has 'an aesthetic and landscape value, and operates within a social organization and a system of shared values' (Laureano 2003). Thus, traditional knowledge operates in perfect harmony with environmental and landscape values, in close compliance with the aesthetic canons set out by tradition. In this sense, the ancient terracing of lemon groves near Sorrento or the gardens of Calabria and Sicily that have enchanted so many visitors for centuries were not conceived of only in terms of productivity, but also as places for delight and recreation.

In our developed societies, none of this remains obvious; on the contrary, the heritage of local knowledge and techniques that followed one another over the centuries has been called into question. In the present-day society, water and its natural cycle are systematically neglected. The result is that water has lost its age old centrality in cultural landscapes that were once built in harmony with ecosystems;

worse, it has become marginal to us. As Eugenio Turri (1998) has argued, the drama of our era is that 'advanced' societies operate on their surrounding environment mainly by favouring 'actors who transform and modify it, leaving behind the mark of their actions', and no longer with 'spectators who can view and understand the sense of their actions', where 'only a human being who feels emotions at the sight of the world … can, above all creatures, find the right path towards building a new and better future.'

Traditional knowledge, in a final analysis, is not so much a *tout court* alternative to modern hydraulic technology, but is rather the necessary condition for its effective and lasting application. Indeed modern technologies, when applied in the abstract – i.e. without taking into consideration history, local know-how and models of water management that developed over the generations – are destined to create at the very least unforeseeable consequences and unsustainable effects.

Today we have considerable means at our disposal to study and revitalize traditional knowledge and the know-how of past Mediterranean water cultures. This knowledge, however, should not simply remain a museum or archival curiosity, but rather become a methodical and interpretative instrument that can be usefully applied to the very problems we face today such as water scarcity, desertification, and flooding. These problems were very often tackled and resolved with a more sustainable vision than any present day approach.

To reinterpret and revalue in modern Mediterranean Italy the inspiring principles of traditional knowledge means – quite unlike the dominant technocratic paradigm – establishing a new model of development aimed at restoring and preserving water, cultural landscapes, and ecosystems of inestimable value with a multifunctional approach founded on its longevity.

Then again, the need to establish a new Water Civilization cannot be separated from the significance of transforming first and foremost the intimate relationship with water in the consciousness of each and every individual, here and now. As Renzo Franzin (2005), the late director of *Civiltà dell'Acqua* once said, 'each existence owes the tone of its voice to the presence or the mirage of waters that have crossed it.' Only with such a 'presence' will water be seen once again as an essential natural and cultural horizon for all action, a gift of the gods, a source of inspiration for artists and poets, a precious Commons that is inextricably part of our environment and our landscape.

References

Altamore, G. 2008. *L'acqua nella storia*. Milano: Sugarco.

Angarano, F. A. 1973. *Vita tradizionale dei contadini e pastori calabresi*. Firenze: Olschki.

Barthes, R. 1957. *Mythologies*. Paris: Plon.

Bayet, J. 1923. Hercule funéraire. *MEFR*, 1 (40), 19–102.

Bevilacqua, P. 1996. *Tra natura e storia. Ambiente, economie, risorse in Italia*. Roma: Donzelli.

Brelich, A. 1964–65. La religione greca in Sicilia. *Kokalos*, X–XI, 35–54.

Caiazzo, A.G. 2003. Simbolismi dell'acqua nell'iniziazione cristiana, in *Storia dell'Acqua*, edited by V. Teti. Roma: Donzelli, 201–224.

Ceravolo, T. 2003. Sacralità dell'acqua, possessione e culto dei santi, in *Storia dell'Acqua*, edited by V. Teti. Roma: Donzelli, 99–112.

Chamoux, F. 1963. *La civilisation grecque*. Paris: Artaud.

Coarelli, F. and Torelli, M. 2000. *Guida archeologica della Sicilia*. Bari: Laterza.

Collin Bouffier, S. 2003. Il culto delle acque nella Sicilia greca: mito o realtà? In *Storia dell'Acqua*, edited by V. Teti. Roma: Donzelli, 44–66.

Cretella, E. 2003. Acque miracolose in Toscana: un percorso simbolico tra religione e magia, in *Storia dell'Acqua*, edited by V. Teti. Roma: Donzelli, 283–292.

Detienne, M. and Vernant J.P. 1979. *La cuisine du sacrifice en pays grec*. Paris: Gallimard.

Dini, V. 1995. *Il potere delle Antiche Madri*. Firenze: Pontecorboli.

Dorsa, V. 1884. *La tradizione greco-latina negli usi e nelle credenze popolari della Calabria Citeriore*. Cosenza.

Eco, U. 1993. *Apocalittici e integrati*. Milano: Bompiani.

Eulisse, E. and Armellin, M. 2008. Prefazione, in *Water, Culture, Society*, edited by E. Eulisse and M. Armellin. Venezia: Centro Civiltà dell'Acqua, 7–10.

Eulisse, E. (ed.) 2008. Saperi tradizionali e gestione sociale dell'acqua nei Sud del Mondo. *Silis, Annali di Civiltà dell'Acqua*, 4–5, 89–231.

Evans, H.B. 1997. *Water Distribution in Ancient Rome: the Evidence of Frontinus*. Ann Arbor: The University of Michigan Press.

Faranda, L. 2003. Simbologia dell'acqua nell'antica Grecia, in *Storia dell'Acqua*, edited by V. Teti. Roma: Donzelli, 151–166.

Franzin, R. 2005, *Il respiro delle acque*. Venezia: Nuova Dimensione.

Ghetti, P.F. 2008. Introduction, in *Water, Culture, Society*, edited by E. Eulisse and M. Armellin. Venezia: Centro Civiltà dell'Acqua, 11–12.

Gilotta, F. 2003. Nota sull'iconografia dell'acqua nel mondo etrusco-italico, in *L'acqua degli dei*, edited by G. Paolucci. Siena: Logo, 25–32.

Guzzo, P.G. 2003. Fonti divine. Miti dell'acqua in Magna Grecia, in *Storia dell'Acqua*, edited by V. Teti. Roma: Donzelli, 35–43.

Hobsbawm, E. and Ranger, T. (eds) 1983. *The Invention of Tradition*. Cambridge: Cambridge University Press.

Jenkins, J.K. 1970. *The Coinage of Gela*. Berlin.

Langill, S. 1999. Indigenous Knowledge: A Resource Kit for Sustainable Development. *The Overstory Agroforestry e-journal*, 82, 35–54.

Laureano, P. 2003. *The Water Atlas. Traditional knowledge to combat desertification*. Matera, Barcelona and Venezia: IPOGEA-LAIA-UNESCO.

Lilliu, G. 1982. *La civiltà nuragica*. Sassari: Delfino.

Lombardi Satriani, L.M. and Meligrana, M. 1989. *Il ponte di San Giacomo. L'ideologia della morte nella società contadina del Sud*. Palermo: Sellerio.

Macbeth, H. and Lightowler, H. 2003. Sete di purezza: l'acqua in bottiglia, in *Storia dell'Acqua*, edited by V. Teti. Roma: Donzelli, 327–344.

Maggiani, A. 2003. Acque "sante" in Etruria, in *L'acqua degli dei*, edited by G. Paolucci. Siena: Logo, 39–44.

Malkin, I. 1987. *Religion and Colonization in Ancient Greece*. Leiden: Brill.

Martinelli, L. 2008. *Piccola guida al consumo critico dell'acqua*. Milano: Terre di Mezzo.

Paolucci, G. 2003. Nel Museo delle Acque: una mostra sull'acqua in età preromana, in *L'acqua degli dei*, edited by G. Paolucci. Siena: Logo, 15–16.

Paolucci, G. and Manconi, D. 2003. Deposito votivo di Grotta Lattaia di Monte Cetona, in *L'acqua degli dei*, edited by G. Paolucci. Siena: Logo, 153–164.

Prigogine, I. and Stengers, I. 1984. *La Nouvelle Alliance. Métamorphose de la science*. Paris: Gallimard.

Pugliese Carratelli, G. (ed.) 2001. *Le lamine d'oro orfiche*. Milano: Adelphi.

Rassu, M. 2004. *Shardana e Filistei in Italia*. Cagliari: Parteolla.

Ravasi, G. 2005. *Le sorgenti di Dio*. Milano: San Paolo.

Reclus, E. 1869. *Histoire d'un Ruisseau*. Paris: Hetzel.

Reij, C., et al. 1996. *Sustaining the Soil. Indigenous Soil and Water Conservation in Africa*. London: Earthscan.

Repellini, F.F. 1989. Tecnologie e macchine, in *Storia di Roma*, edited by A. Momigliano and S. Schiavone. Torino: Einaudi, 323–368.

Ricciardi, L. 2003. Deposito votivo del Santuario di Fontanile di Legnisina di Vulci, in *L'acqua degli dei*, edited by G. Paolucci. Siena: Logo, 125–132.

Rodgers, R.H. 2004. *Frontinus: De aquaeductu urbis Romae*. Cambridge: Cambridge University Press.

Ruf, T. and Valony, M. 2007. Les contradictions de la gestion intégrée des ressources en eau dans l'agriculture irriguée méditerranéenne. *Cahiers Agricultures*, 16 (4), 294–300.

Scarborough, J. 1969. *Roman Medicine*, London (repr. 1979. New York: Ithaca).

Scheid, J. 1989. Religione e società, in *Storia di Roma*, edited by A. Momigliano and S. Schiavone. Torino: Einaudi, 631–660.

Sébillot, P. 1983. *Les eaux douces*. Paris: Imago.

Shiva, V. 2002. *Water Wars. Privatisation, Pollution and Profit*. Cambridge MA: South End Press.

Tamino, G. et al. 2002. *Etica, biodiversità, biotecnologie, emergenze ambientali*. Padova: Isonomia.

Teti, V. 2003. Luoghi, culti, memorie dell'acqua, in *Storia dell'Acqua*, edited by V. Teti. Roma: Donzelli, 3–34.

Tölle-Kastenbein, R. 1990. *Antike Wasserkultur*. Munich: Verlag Beck.

Trevor Hodge, A. 1992. *Roman aqueducts and water supply*. London: Duckworth.

Turri, E. 1998. *Il paesaggio come teatro*. Venezia: Marsilio.

UNCCD 2005. *Promotion of Traditional Knowledge*. Bonn: UNCCD.

Vallerani, F. and Varotto, M. 2005. *Il grigio oltre le siepi*. Venezia: Nuova Dimensione.
Vegetti, M. and Manuli, P. 1989. La medicina e l'igiene, in *Storia di Roma*, edited by A. Momigliano and S. Schiavone. Torino: Einaudi, 389–432.
Vitali, D. et al. 2003. Il deposito votivo di Monte Bibele, in *L'acqua degli dei*, edited by G. Paolucci. Siena: Logo, 111–120.

PART II
Law, War, and Water

Chapter 4
The Right to Have Water: Or an Obligation to Provide It?

Keith Porter

Introduction

On 30 April 2009, a news report described how Mexican health authorities were advising citizens to wash their hands frequently to protect against the threat of swine flu. Unfortunately, as the reporter pointed out, water is a luxury for many in Mexico, even to wash hands (Beaubien 2009). This story dramatically demonstrates the reality of the critical scarcity of water experienced by a large fraction of the world's population.

In an International Forum held at Cornell University in March 2009,[1] speakers addressed the shortage of water in countries in the Mediterranean region. One speaker stated that in Greece, for example, the shortage is as much due to corruption and bad management as it is a consequence of a hydrologically based scarcity of water. However, in African countries lying within the hydrological system of the Mediterranean region, such as in the Nile Basin, there is clearly an absolute scarcity of water available for the use of their citizens. This chapter addresses the latter condition as posing the most vital challenge facing the welfare of the people of the Mediterranean region and indeed the entire world.

Many Mediterranean countries, especially the north African nations, face a potentially calamitous water crisis. The gap between the demand for water and the water available to meet that demand is increasingly a chasm. In consequence, public health, economic development, ecosystems and bio-diversity in the region are all seriously jeopardized. Also potential conflicts over water shortages threaten peace and stability in the region. As a recent Mediterranean Civil Society Meeting (2008) concluded, having a strategy for sustainable water management in the region is imperative. This strategy should acknowledge adequate water is a prerequisite for human, social, economic and cultural development. Adequate water resources are also essential for sustaining vital natural and biological resources. The Mediterranean Civil Society meeting further declared that *Governments cannot face this crisis alone*. Given the comprehensive reach of the water needs, it seems necessary that principal stakeholders must play their part in dealing with the crisis.

---

1 Cornell International Forum: 'Water-sharing and Culture in the Mediterranean.' A. D. White House, 6–8 March 2009.

Grass roots movements from all major groups, women, youth, indigenous peoples, unions, water users, farmers, local authorities, science and technology, business and NGOs have critical roles as partners with government's efforts to achieve sustainable management of their water resources.

The Problem of Water Scarcity

It is universally accepted that the availability of adequate and affordable water is a pre-condition to a reasonable quality of human life and economic vitality. '[W]ater is used in virtually everything we make and do. It is the most widely used resource by industry; it is used both directly and indirectly to produce energy; it provides the basis for much of our outdoor recreation; it is an important part of our transportation network; it serves as a vehicle for disposing of wastes; and it provides important cultural and amenity values'(Frederick 1995). 'Virtual water' is a measure of this statement. Introduced and expounded by Professor Allen, 'Virtual water,' is a construct that quantifies the amount of water embedded in and which is necessary for products to be created (Allen 1998). Two common examples are apples and paper. The estimated virtual water content of one apple is seventy liters, and that estimated for a single sheet of A4 paper is ten liters.[2] A recent report estimates that the average Briton consumes 4,645 liters of water a day when virtual water is added to personal use of water. Even people in poor countries 'typically subsist on 1,000 litres of virtual water a day' (Lawrence 2008).

Given this total dependence upon water, population growth, economic development, and now climate change triply burden the world's water resources. Remarkably, there is a 'lack of a comprehensive treatment of water issues in international law' (Cullet and Gowlland-Gualtieri 2005). An increasingly urgent question is whether the escalating demand for water can be sustained by its supply. United Nations Secretary General Kofi Annan has noted that '[o]ne person in six lives without regular access to safe drinking water.'[3] This figure is worsening.[4] This scarcity of water as a vital necessity for human life poses increasingly asked key questions: Is there a human right to water, or is it more desirable to consider the obligation to remedy critical shortages of water?

The first part of this chapter briefly reviews the progress of the idea of a human right to water and considers the potential advantage of focusing on the obligation posed by human needs as opposed to rights. The latter part of the chapter suggests

2 Waterfootprint.org, Product Gallery, http://www.waterfootprint.org/?page=files/productgallery (last visited 9 January 2008).

3 Press Release, Secretary General, (2003).

4 See, e.g., Marianne Lavelle and Joshua Kurlantzick, 'The Coming Water Crisis,' US News & World Rep., 4 August 2002 (describing the water crisis in America); United Nations Environmental Programme, World Environment Day – 5 June 2003, available at http://www.unep.org/wed/2003/keyfacts.htm.

that the focus on the question of human rights to water, or indeed other economic and social rights, has lost sight of the fundamental need to address the basic welfare of society. Without relinquishing the gains made by the Human Rights Movement experienced since the Second World War, it is desirable to revive the basic concern for social welfare as the context for the human right to water.

An Ancient Right and Obligation

If water is scarce, humans are more likely to be thirsty. It is therefore understandable that a right to water or obligations to provide it would evolve where water is lacking. This is true of the Middle East where the dry climate does not appear to have changed since the first known settlements there (Beitzel 2007). The 'right of thirst' or the obligation to satisfy it accordingly can claim an ancient history extending back to the original sources of the Jewish, Christian, and Islamic religions. As the Bible instructs: give one's enemy water to drink if he is thirsty.[5] Talmudic law expands upon that instruction by recognizing the right of every traveler to use a drinking water well (Caponera 1992). Islamic law similarly recognizes the human necessity 'to take water to quench one's thirst or to water one's animals' (Caponera 1973). According to the Sunnah, 'To the man who refuses his surplus water, Allah will say: 'Today, I refuse thee my favour, just as thou refused the surplus of something that thou hadst not made thyself' (Caponera 1973). A further statement of the Sunnah tradition is that '[n]o one can refuse surplus water without sinning against Allah and against man' (Caponera 1973). Despite being more thoroughly affirmed in the Sunni tradition, Sunni and Shi'ite doctrine alike recognize this Right of Thirst (Hirsch 1959). More recent articulations of Islamic law, including the Ottoman Empire's civil code – the Mejelle, affirm the basic water right of everyone to quench his or her thirst from any river (Caponera 1992).

Despite this attestation of the obligation to satisfy the necessity for water in religious laws originating in the Middle East, later recognition of a human right to water in international law has remained cautious. There are few explicit, legally binding recognitions of a human right to water. Under international humanitarian law, the Geneva Convention of 1949 explicitly affirms that sufficient potable water and water for personal washing shall be supplied to prisoners of war. Further, the Additional Protocols I and II of the Geneva Convention prohibit in any way damaging the drinking water installations serving local civilian populations. The Convention and its Protocols thus clearly recognize a right to water among those engaged in or affected by warfare.

Other legally binding instruments have only implicitly recognized a right to water. The United Nations Charter Article 55 promotes social progress, development and health. This declaration was given legal expression in the International Covenant on Economic, Social and Cultural Rights which recognizes everyone's

5 Proverbs 25: 21.

right to a sufficient standard of living including adequate food and hygiene. No explicit mention is made of water however. This omission was remedied in part by a Convention recognizing rights of women in 1979 to include adequate living conditions 'particularly in relation to housing, sanitation, electricity and water supply.'[6] Ten years later, the Convention on the Rights of the Child affirmed the right of the child to the enjoyment of the highest attainable standard of health including clean drinking water.[7] However, these are the only direct statements affirming a right to water for a segment of populations not involved in a war. Conversely, therefore, it may be argued that an adult male who is not a prisoner of war, nor a civilian in a militarily occupied territory, does not have a legally enforceable right to water under international law.

There are key non-binding instruments that advocate a human right to water. The Universal Declaration of Human Rights (UDHR), which has substantially influenced customary international law concerning human rights notes that 'a standard of living adequate for the health and well-being of himself and of his family . . . include[es] food, clothing, housing and medical care and necessary social services.'[8] Since the factors listed are inclusive and not exhaustive, and a standard of living adequate for health is impossible without water, this right may be interpreted as necessarily including water (McCaffrey 2005). It may be noted that two years before the UDHR, the World Health Organization (WHO) affirmed an explicit right to good health in its constitution signed on 22 July 1946. The Preamble to the Constitution states: 'The enjoyment of the highest attainable standard of health is one of the fundamental rights of every human being without distinction of race, religion, political belief, economic or social condition.' Such a high standard of health obviously mandates the provision of high quality water.

Other non-binding instruments have referred explicitly to the human right to water. For example, Guiding Principle No. 4 of the 1992 Dublin Statement on Water and Sustainable Development states it is vital to recognize the 'basic right of all human beings to have access to clean water and sanitation at an affordable price.' A possibly more influential instrument is a comment issued by the Committee on Economic, Social and Cultural Rights. In 2002, the Committee issued its General Comment No. 15 which states '[t]he human right to water is indispensable for leading a life in human dignity. It is a prerequisite for the realization of other human rights.'

---

6 United Nations Convention on the Elimination of All Forms of Discrimination against Women, art. 14(2)(h), 18 December 1979, 1249 U.N.T.S. 14.

7 Convention on the Rights of the Child, art. 24(2)(a), art. 1, 20 November 1989, 1577 U.N.T.S. 3.

8 Universal Declaration of Human Rights, Article 25(1).

The Key Instrument

The key international instrument regarding an assumed human right to water remains the Covenant on Economic, Social and Cultural Rights. Unfortunately, even if the Covenant can be interpreted as advocating the human right to water it does not greatly assist in meeting that right. The covenant embodies several impediments to favoring obligations (Sepulveda 2003). First, the language of the covenant is general and vague. It does not provide for any procedure for individual petitions. Further, and crucially, the social rights do not reflect domestic legal norms, as did the Covenant for civil and political rights (Alston 1992). Lauterpacht has commented on this difficulty posed by the category of rights that 'for the sake of convenience may be referred to as social and economic rights' (Lauterpacht 1968). Traditional bills of rights do not prominently include such claims. However, as Lauterpacht points out, 'the precious rights of personal liberty and political freedom may become a hollow mockery for those whom the existing social and economic order leaves starving, insecure in their livelihood, illiterate, and deprived of their just share in the progress and well-being of the society as a whole'. It follows that International Bills of Rights are incomplete and fatal to their authority to the extent human social and economic claims are left out of account. Nevertheless social and economic rights are not enforceable as are political and civil rights. Rather than being enforceable under constitutions, they are positive claims established by specific legislation. Under international law, social and economic rights 'can be made real inasmuch as they will constitute a legal obligation of the State – an obligation subject to international interest, to discussion and recommendation by international agencies and, in case of grave and persistent violations and neglect, to appropriate international action' (Lauterpacht 1968). It can certainly be argued that the right to water is a hollow mockery for those who are thirsty in a land that lacks the means to remedy that thirst.

General Comment No. 15

The UN Committee on Economic, Social and Cultural Rights has tried to address such considerations in its General Comment No. 15: The Right to Water. Substantively, the General Comment suggests that the right to water encompasses three requirements. The water must be available with adequate quality and in sufficient quantity. It must also be accessible. Accessibility also entails the necessity that the water be affordable. As the Comment explains:

> Article 11, paragraph 1, of the Covenant [on Economic, Social and Cultural Rights] specifies a number of rights emanating from, and indispensable for, the realization of the right to an adequate standard of living 'including adequate food, clothing and housing'. The use of the word 'including,' indicates that this catalogue of rights was not intended to be exhaustive. The right to water

> clearly falls within the category of guarantees essential for securing an adequate standard of living, particularly since it is one of the most fundamental conditions for survival.

In part III the Comment sets out general, specific, international and what it terms core legal obligations. The general legal obligation is for States parties to move expeditiously and effectively towards the full realization of the right to water. The right to water creates three specific obligations: to *respect,* to *protect,* and to *fulfil*. The obligation to *fulfil* in particular is an obligation to take positive measures to facilitate promote and provide a national water resources strategy and plan of action that ensures affordable water for all. Such a strategy importantly should ensure 'there is sufficient and safe water for present and future generations'. Thus the strategy to satisfy the right to water should be sustainable. The obligation to *fulfil* also specifically includes access to adequate sanitation that is fundamental for human dignity. Also, most importantly, sound sanitation is a primary means of protecting the wholesomeness of drinking water supplies.

The international obligations are largely negative in urging that states parties and their citizens refrain from jeopardizing the security of water supplies of other countries. 'Any activities undertaken within the State party's jurisdiction should not deprive another country of the ability to realize the right to water for persons in its jurisdiction.' This accords with the international principle not to cause Transboundary harm.[9] A more positive international obligation is to ensure that the right of water is given due attention in international agreements and organizations relevant to the management of water.

In its General Comment, the Committee also identified what it termed as core obligations. Of the nine core obligations, the first four ensure that access to safe water is achieved sufficiently to prevent disease, and is provided on a non-discriminatory basis, without excessive expenditure of effort or risk to personal security. It is also a core obligation to ensure equitable distribution of water services. A national water strategy should be adopted and implemented in a manner that is participatory and transparent. Realization or otherwise of satisfying the right to water should then be monitored. Vulnerable and marginalized groups should be especially protected in ensuring that their right to water is realized. Finally, the ninth core obligation acknowledges the right to water in its context of public health by specifying that, 'measures should be taken to prevent, treat and control diseases linked to water'.

From his examination of the General comment 15, McCaffery draws several conclusions (McCaffrey 2005). First, he suggests it is tautologous to conclude that water is essential for other human rights established by the covenant because without water there is no life and other rights are simply null. However, General Comment 15 as a basis for advancing the right to water is limited as a legal

9 See the Rio Declaration (UN Conference on Environment and Development, 1992), Principle 2.

instrument. The Committee has no law making powers or authority. Therefore the standing and ultimate influence of its General comment depends upon its acceptance or adoption by states. Such adoption should solicit corresponding obligations.

The General Comment does serve a purpose in drawing attention to the critical problem of the scarcity of water and in advocating steps that can be taken by states to deal with that scarcity. With respect to satisfying the right to water, the General Comment recognizes that the right cannot be realized immediately but that states have an obligation to take steps on a non-discriminatory basis. A less positive aspect of the General comment, McCaffery suggests, is that the scope of its nine core obligations could 'do more harm than good'. Their scope and range may be unrealistic and raise overly high expectations. However, he commends the Committee for considering the international aspects of the right to water. Increasing inter-dependence between nations regarding shared water resources will necessitate international understandings and agreements. River basins that are shared by two or more countries, account for nearly half of the total land surface of the Earth (UNEP 2002). In the Mediterranean hydrological system, a most important example of such an international river basin is the Nile River basin shared by 10 nations.

Returning to Obligations?

As a United Nations report has observed, a human right is inseparable from considerations of its implementation. For example, the formal recognition of the human right to water immediately raises many practical questions. These include the critical questions: Who has responsibility for satisfying the obligation to obtain and supply the water? From what sources is the water to be supplied? How is the supply of water to be accomplished?

Meeting the Obligation to Ensure Adequate Water

Where water is scarce there are multiple options by which such scarcity could be remedied at least in theory. The governing calculation lies in the balance between the supply of water and its consumption. If the scarcity of water is to be remedied, either the supply must be increased or the demand for water decreased. An obvious way to reduce the demand for water is to substitute uses that require less water than do current uses. An example would be for farmers to change their crop production from plants that require large amounts of water that is largely lost via evapo-transpiration to crops which are more frugal in their water needs. Another example would be changing from water consumptive farming to industrial uses that use water more sparingly. A related option that can effectively reduce demand is to match the quality of water consumed to the requirements of the use.

In 1958, the United Nations Economic and Social Council stated: 'No higher quality water, unless there is a surplus of it, should be used for a purpose that can tolerate a lower grade [of water]' (United Nations 1958).

Regarding the supply of water, conservation and recycling of water may be an option that increases the amount available to its users. This option is equivalent in its effect to increases in supply by re-using the water. Assuming all existing water resources in a water scarce region have been exploited, then obviously the only way to increase the supply is to develop a new supply such as by desalination. Transfer of water from a region with surplus water to a water scarce region also would alleviate the scarcity of water in the latter. In the Mediterranean region, Cyprus is an example of a country where the scarcity of water is compelling the island to find ways to increase its supply of water. This is illustrated by the plans or arrangements to provide water to Cyprus from Greece and Turkey. Cyprus also plans to construct a desalinization plant at Episkopi on the island.[10]

An important consideration is the geographic scale defined by the transfer. The scale may be local, regional, national or international. This has consequences for the level of governance that is required. According to the size and extent of the region within which the solutions apply, the governance can vary from local management to the requirement for international agreements. Regardless of the geographical scale involved, transfers of water would reduce the potential for overexploitation of existing water resources by excess demand in the receiving region that otherwise would likely be unavoidable. Sustained overexploitation of water resources can lead to their irreversible damage and destruction.[11] It also follows the rights and needs of those in the supplying region must be taken into account. It is foreseeable that a state could potentially transfer water in a way that neglects the interests of its own water poor citizens. Such a situation may be anticipated where the demand for water in the receiving region represents a higher valued use of the water than that existing in the supplying region.

The demand for water comprises its multiple human uses. Likewise managing and developing new supplies should take into account those affected by the diminishment of the resource on the one hand and the gain in supply on the other. Both sets of considerations may involve the interests of stakeholders at personal, community and institutional levels. Being short of water has significant negative effects upon a locality's or a region's economic development.[12] A recent report

---

10 *Global Water Intelligence* (March 2009). 'Episkopi award in the lap of the gods.' See http://www.globalwaterintel.com/archive/10/3/general/episkopi-award-in-the-lap-of-the-gods.html.

11 See Robert Glennon, *Water Follies: Groundwater Pumping and the Fate of America's Fresh Waters* 23–34 (2002) (describing the effects of the over-pumping of groundwater). See also supra sec. I.

12 Cf. Hsiang-Chih Hwang, 'Water Quality and the End of Communism: Does a Regime Change Lead to a Cleaner Environment?' 28, available at http://papers.ssrn.com/sol3/papers.cfm?abstract_id=952320 (last visited 22 March 2008) (concluding that 'regime

noted that among the world's poor countries, those with access to improved water and sanitation services experience greater economic growth than those without improved services.[13] This fundamental worth of water prompts a key question: To what extent can stakeholders and local communities participate in the decision making about water resources, which directly concerns their own economic interests, but which is customarily the province of higher levels of government? Planning and management of water resources is typically a 'top-down' affair.

Aarhus Convention: Providing a Critical Role for Stakeholders?

Scholarly discussions of human rights generally, and of the right to water in particular, also tend to focus on a 'top down' perspective. However, the use and management of water is generally local in character (Porter 2005). Even the provision and transfer of water on a large scale can raise local issues. Therefore it is appropriate to consider how the human right to water of those in the supplying or receiving regions may participate in the consideration of options and the subsequent decision-making. Such a shared role in making and implementing decisions could be supported by the application of the Aarhus Convention.[14] This convention was 'an important milestone in terms of international legally binding instruments.'[15] The Aarhus Convention establishes the principle that sustainable development requires the involvement of all stakeholders. Such involvement is based on the obligation of states and public authorities to provide the general population with access to information, participation in the decision-making and access to administrative and judicial justice.

Accordingly, the Aarhus Convention could protect the interests of those who receive water supplies, and also those of the residents in the region from which the water is supplied. In making available full information to stakeholders about the sale of what they may regard as *their water*, providing for their involvement in the decisions regarding water resources development, diversions or transfers,

---

change [in post-communist Eastern European countries] had a favorable impact on water quality except in the case of dissolved oxygen, which may reflect the effects of illegal oil discharges.')

13 Stockholm International Water Institute, 'Making Water a Part of Economic Development' 4 (2004), available at http://www.siwi.org/downloads/Reports/CSD_Economics.pdf (last visited 7 January 2008) ('Poor countries with improved access to clean water and sanitation enjoyed annual average growth of 3.7 per cent. Poor countries with the same per capita income but without improved access had an average annual per capita GDP growth of only 0.1 per cent.')

14 Convention on Access to Information, Public Participation in Decision-Making and Access to Justice in Environmental Matters, 25 June 1998, 2161 U.N.T.S. 447 [hereinafter Aarhus Convention].

15 The Environment for Europe Process, The Aarhus Environment for Europe Conference, http://www.environmentforeurope.org/efehistory/aarhus1998.html (last visited 16 August 2008).

and uses of water, and providing for administrative or judicial procedures in the event of stakeholder objections, decision-makers could be held accountable by their citizens. Such requirements of the Aarhus Convention could also protect third party interests invoked by the uses of water or by water transactions.[16] The Aarhus Convention therefore appears to be an effective international instrument that would prompt States to satisfy their core obligations to 'ensure access to the minimum essential amount of water, that is sufficient and safe for personal and domestic uses' (Committee on Economic, Social and Cultural Rights 2002).

Putting Rights into their Context

A great deal of ink has been expended discussing whether or not there is a Human Right to Water in International Law. While the worthiness of the debate occasioned by this expenditure is unquestionable, its value in diminishing thirst for water in the world can also be debated. The term Human Right to Water has also been much 'trampled' on in scholarly writings.

There are three objections to the debate over whether or not there is an international human right to water. First, the international declarations themselves in affirming human rights in their various guises may serve a diversionary rather than positive role in advancing the Rights. In his book *Crimes Against Humanity* Robertson (2006) describes the conduct of United Nations Committees that meet, worry about the nuances of language they consider, and exercise caution over the words they eventually adopt. By such linguistic efforts they successfully avoid commitments and enforcement that would otherwise give meaning to the words they have so carefully composed. In so focusing on the issue of Rights themselves, the debate also neglects to consider explicitly how wholesome water can and should be provided to people lacking adequate water to ensure their 'rights to that water' are in fact satisfied.

The Context of Social Welfare

Satisfaction of the need for water also generally must occur in the context of other pressing economic and social human needs, including the basic necessities for food, clothing or sound housing. Generally the need for water is not an isolated need. The necessity for water is frequently accompanied by other necessities. These may be viewed as a bundle of sticks that together provide for a satisfactory basic level of life. Sidney and Beatrice Webb represented the insufficiency of

16 See generally A. Dan Tarlock, 'Water Transfers: A Mean to Achieve Sustainable Water Use,' in *Fresh Water and International Economic Law* 54–58 (Edith Brown Weiss, Laurence Boisson De Chazournes and Nathalie Bernasconi-Osterwalder eds, 2005) (describing third party interests in water transactions in the Western United States).

necessaries of life to the extent that health and life were put at risk as destitution or a *disease of society* (Webb 1911). The critical need for water suffered by most people in the world today is an aspect of their destitution. With this setting, the right to water becomes a matter of individual or social welfare.

As McCaffery (2005) asserts: 'The human right to water is crucial to the welfare of people around the world'. By nurturing that right 'we who inhabit Earth today can discharge our obligations to succeeding generations'. Welfare of the people is in practice the purpose of social rights. However, it may be argued that an undue current focus on human rights loses sight of our primary responsibility to protect the health and welfare of people. That focus can too readily slide into substituting conceptualism for practice.

Unhappily, the multiple conceptual maps that have been produced to represent it have blurred the notion of 'Welfare'. Indeed it is suggested the word welfare 'identifies a particularly contested part of the conceptual landscape that has been much trampled by economists, philosophers and political theorists, as well as a wide variety of more practical politicians, policy analysts and social commentators' (Hamlin 1993). In political contexts the word welfare has especially acquired conceptual baggage that is unhelpful. Despite this handicap of language, economic and social welfare historically remains a sound objective.

For the purposes of this chapter, the simple dictionary definition of welfare will be applied: 'In the broadest sense, the good of the people. The well being of an individual. In a more limited sense familiar in modern usage, public relief for the poor, as provided under relief or welfare acts' (Ballentine 1969).

Providing for Social Welfare

What is the historical basis of the recognition of the consideration due one's fellow men and women? At least in the Western world, an active concern for the welfare of the poor or destitute has a long history. Early Christianity had a strong notion of stewardship with respect to those in need. 'To withhold alms when there is evident and urgent necessity is mortal sin.'[17] Throughout Christendom in Europe, the mediaeval Church, benevolent institutions including Craft Gilds, Monasteries and municipal corporations all provided for the poor and needy. However, as Sidney and Beatrice Webb describe (1911), in the first quarter of the 16th century there was a significant departure from this tradition of assisting those in need. Rather than the former reliance on the emotion of pity and Christian charity it was increasingly seen as in the public interest to provide for the destitute through civil administration. It is apparent, 'that the movement for taking the task out of the hands of the church, and dealing with it as a part of civil government, was common to practically the whole of Europe' (Webb 1911). The consequence

17 St. Thomas Aquinas, *Summa Theolg* 2a2ae, Q. xxxii, art. V. (Cited in: R.H. Tawney, Religion and the rise of Capitalism 260 (Peter Smith 1962) (1926).

was a secularization of relief for the poor and needy increasingly divorced from any latent notion of their rights. Instead, the emphasis was on meeting a social obligation.

By the 19th century industrial development elevated the question of social needs to a new level of urgency. This was especially the case in England, which was in the forefront of industrial development. Appalling social conditions were created in that country by the rapid urban development as the industrial revolution progressed. Although not well understood at the time, the growing problems of the lower classes in particular prompted an increasing recognition of the urgent need for remedies. 'The growth of an industrial-urban society enormously magnified and altered the problems of the common life, and these called forth gallant, often sacrificial, effort on the part of individuals and groups of Englishmen' (Owen 1964). In response, private individuals and benevolent groups provided aid for hospitals, schools and the poor. It may therefore be claimed, 'philanthropists contributed something – in some instances a good deal – toward relieving tensions in English life' (Owen 1964).

Despite this resumption of a form of private almsgiving to assist the needy, it is relevant to note that responses to the social crisis were not generally initiated under the guise of defending human rights. In fact evidence suggests that there was little humanitarian motivation for the persistent attempts to remedy the worst consequences of the industrial revolution. Rather there was an accepted sense of social and economic needs with which civil administration was obliged to deal. Thus the problems were regarded as being a matter of obligations rather than in terms of social rights. With respect to the problems of water supplies and human waste, the outcome was the so-called Sanitary Movement.

The Sanitary Movement

The Sanitary Movement has a double relevance to the question of the human right to water. First, as indicated above, the Sanitary Movement was based on a sense of governmental obligation rather than on any ideals of social or human rights. Secondly, developing nations, where the scarcity of water is a vital concern, can experience economic growth that mirrors in varying degrees the experience of developed countries in the 19th century. Initial and rapid economic development frequently comes at the expense of the welfare of the working and poorer sectors of the population.

This is especially the case with water. During the 19th century, European and North American inhabitants of the rapidly industrializing urban centers suffered abominable living conditions including the lack of wholesome water. The social distress and economic cost of high mortality and very poor health motivated the 'Sanitary' idea. Ann La Berge (1992) suggests that France was first in addressing the needs of public health. As the 19th century progressed however, England increasingly was in the forefront in promoting new public health measures. Melosi

(2000) has termed this ascendancy of British Public Health 'a triumph'. Much credit for this accomplishment is owed to Edwin Chadwick (1842). Chadwick was a barrister by profession who had a highly influential, if unpopular, career in public service promoting ideas of wholesome drinking water, sanitation and public health. Upon his death, *The Times* famously declared 'We would rather take our chances with cholera than be bullied into health by the likes of Mr. Chadwick.' Nevertheless, 'the nation owes an immense debt to him' (Binnie 1993). In the latter half of the century public officials and water engineers both in the UK and in the US increasingly adopted Chadwick's ideas. Nevertheless, this victory should not be considered a triumph for human rights. As Rosen states, the success of the sanitary idea was not simply due to humanitarian sentiment or social conscience. Rather it was fostered by the recognition of the social and economic costs of poor water supplies and sanitation. As John Simon declared in 1858, 'Sanitary neglect is mistaken parsimony' (Rosen 1993). In other words, a principal motive for providing safe drinking water and sanitation is economic rather than humanitarian.

20th Century Advance of Human Rights

In the 20th century the focus shifted towards explicit consideration of human rights. Following the First World War, Lauren (2003) suggests a striking feature of the period was the 'dramatic increase' in the extent of discussions on human rights around the world. Given the abounding international animosities and the opposition of nations to interference with their sovereignty, it is surprising what was accomplished in the early decades of the century. These accomplishments included the promotion of the rights of women and children, and the adoption of the Geneva Protocol signed on 17 June 1925,[18] and the Convention relative to the Treatment of Prisoners of War, Geneva, 27 July 1929.

H.G. Wells Campaign for Human Rights

These successes were nevertheless very limited in number and scope. Indeed, it is suggested that one of the great mysteries of the 20th century was the virtual silence on the part of European intellectuals and politicians concerning human rights. This silence was resolutely maintained in the face of the extermination of millions of innocent people by the Soviets and the increasingly vicious mistreatment of Jews in Germany (Robertson 2006). During and certainly by the end of the Second World War however, there was an absolute and resolute condemnation of the holocaust. The affirmation of human rights during and after the Second World War is largely attributed to this condemnation. In this response to the atrocities of the Nazis, the role of H.G. Wells was highly influential although today generally

18 The protocol was ratified by many countries, including major powers except the United States and Japan. The United States ratified the protocol on 22 January 1975.

overlooked. Robertson describes how the 'revival of the human rights idea in the 20th century really began at the instigation and inspiration of the British author H.G. Wells.' (Robertson 2006). In two letters he wrote to *The Times* at the beginning of the Second World War, H.G. Wells advocated the adoption of a declaration of rights. His campaign quickly attracted 'extraordinary support in England' (Robertson 2006). Wells' campaign gained further influential force through the publication of a Penguin Special, *H.G. Wells on the Rights of Man* (1940). The book was an immediate success, selling thousands of copies. It was translated into thirty languages. Newspapers throughout the world syndicated the publication. Especially important was its influence on President Franklin Roosevelt who was a friend of Wells. Ideas from Wells' booklet were incorporated into Roosevelt's appeal in 1941 for a world based on the four essential freedoms of speech and worship, and freedom from want and fear.

In his book, Wells claimed: 'To a large extent the obligation to provide subsistence, shelter and care is already recognized in practice in the more civilized regions of the earth' (Wells 1940). Among other rights therefore, it is expedient for a Declaration of Rights to declare: 'That every man is joint heir to all the resources, powers, inventions and possibilities accumulated by our forerunners, and entitled without distinction of race, colour or professed belief or opinions, to the nourishment, covering, medical care and attention needed to realize his full possibilities of physical and mental development and to keep him in a state of health from his birth to death' (Wells 1940).

Welfare Rights Gaining Ground?

Such ideas of welfare rights especially gained ground in the UK and the US during the 1960s (Cane and Conaghan 2008). T.H. Marshall was especially influential. In his seminal essay, 'Citizenship and Social Class,' he argued that citizenship consists of a bundle of political, social and civil rights (Marshall 1950). A universal welfare scheme then arises from these rights. Unfortunately, with respect to welfare rights, as James Nickel (2007) points out: 'The fact that some countries are rich and others are shockingly poor makes uniform worldwide standards problematic'. Such disparity in wealth makes uniform accomplishment internationally also virtually impossible.

With respect to the international context, the formative event promoting welfare rights for the world's population was the Universal Declaration of Human Rights (UDHR). Robertson asserts a fundamental criticism of the Declaration: 'States voting in favor of the Universal Declaration in 1948 did not anticipate for a moment that their vote meant they were assuming any obligations to enforce the rights declared'. Despite this pessimism over the precept versus its practice, it is claimed that the UDHR has had key significance at national and international levels extending to moral, political and legal spheres (Eide and Alfredsson 1992). In particular, Article 25 of the UDHR affirms:

Everyone has the right to a standard of living adequate for the health and well-being of himself and his family, including food, clothing, housing and medical care and necessary social services, and the right to security in the event of unemployment, sickness, disability, widowhood, old age or other lack of livelihood in circumstances beyond his control.

Article 25 is a key component in affirming economic and social rights. Its comprehensive articulation of such social rights constitutes a welfare scheme. As often noted however, despite the comprehensiveness of these welfare rights the right to water is not specifically mentioned. In 1966, the UN General Assembly adopted the International Covenant on Economic, Social and cultural Rights affirming the welfare rights in the UDHR. In its specification of rights, the Covenant also omitted the right to water. However, without water there is no life. Therefore, since the identified social or welfare rights would be of no account without water, it seems safe to assume the right is implied. The omission may also be taken to support the idea of considering the right to water as an essential part of the bundle of social rights articulated in the Covenant.

With this view it follows that pursuing the objective of ensuring adequate water for all humans requires simultaneous consideration of their overall welfare. In this objective there would be a revival of the emphasis on social obligation as demonstrated in the 19th century by the likes of Edwin Chadwick.

Conclusions

As the UN Committee on Economic, Social and Cultural Rights affirms, water 'is a limited natural resource'. It is also 'a public good fundamental for life and health'. Regardless of this fundamental necessity for human life, 'Human rights were not a free gift. They were only won by long, hard struggle' (Lauren 2003). Despite this confident claim of UNESCO, the reality is less positive. Human rights have not been won. A specifically clear demonstration of the failure is the right of humans to water. 'By 2025, 1,800 million people will be living in countries or regions with absolute water scarcity, and two-thirds of the world population could be under stress conditions.'[19] To assure the woman dying of thirst she is a winner because she has a right to water is little consolation to her as she dies.

The 'right to water' is a so-called 'claim' right in that there is a corollary duty for someone to satisfy the right (Cane and Conaghan 2008). As such it is a right included among the class of human rights 'that morally ought to be secured for all human beings.' To date, international law has escaped this moral duty because the human right to water is not yet treated as a legally enforceable obligation. Some scholars have suggested that international law has not explicitly recognized a binding human right to water because any such right

19 http://lowband.fao.org/lbp?http://www.fao.org/nr/water/issues/scarcity.html (Last read 27 March 2009).

would be unenforceable. For many nations where the need for water is greatest, their ability to provide water for their populations can be almost entirely beyond their control. Their capacity is constrained by the very low standard of living sustained by their economies. The character of a nation's water resources is also a function of its hydro-geography. That character may not provide options for the acquisition of additional supplies of water without investments beyond the means of the nation. In such circumstances to insist on a human right that is beyond the capacity of those upon whom the duty falls and is hence unattainable simply discredits the right. The necessary conclusion is that, rather than debate whether or not a right to water exists, it would better serve to concentrate on how the capacities of states might be increased to meet their obligations to provide water to those in need of it.

The focus on economic and social rights as opposed to obligations became the practice, particularly following the Second World War. In Europe, the concern for those in need of the necessaries of life was originally a matter of Christian charity. During the reformation there was a change in favor of municipal units accepting the responsibility for the destitute as a matter of public interest. This primary role of government continued through the 19th century and was particularly the case as society tried to meet the evils of industrialization. A particular evil was the unwholesome condition of water supplies that caused calamitous outbreaks of water-borne diseases in Europe and the United States. Through the inspired leadership of men such as Edwin Chadwick, the Sanitary Movement was developed to provide and protect drinking water. This movement was based on governmental responsibilities assumed with an increasing sense of duty.

In the 20th century there was tendency to focus on the enunciation and declaration of human rights. This was especially a response to the Holocaust and other acts of genocide. As the human rights movement has matured however, it has become increasingly timely to give more attention to the role of governments in meeting the obligations corresponding to the human rights. It is abundantly clear we live in a world of economic limits. This reality applies to the African and Middle Eastern countries comprising the eastern and southern regions of the Mediterranean. Given the constraints of these economically developing nations, the capacity to meet social rights could be strengthened by engaging the resources of NGOs, corporations and public groups. Agenda 21 states: 'One of the fundamental prerequisites for the achievement of sustainable development is broad public participation in decision-making' (Robinson 1993). This is an aim of the Aarhus Convention. The goal of public participation in its broadest sense could more meaningfully answer the questions: To what extent is a legal human right to water practicable? How can recognition of such a right at the local level encourage participation in decision-making necessary to remedy scarcities of water? What legal instruments and other technical capacities may assist or protect the interests of stakeholders regarding development of in situ water resources, potential intra and inter-basin diversions, and the bulk international transfers of water? How can an effective partnership be established between all levels of government to establish sustainable solutions to the problem of water scarcities especially in developing countries?

According to the Johannesburg Declaration on Sustainable Development (2002), there are three 'interdependent and mutually reinforcing pillars of sustainable development – economic development, social development and environmental protection.' However, without adequate water for human, social and economic needs, none of these pillars stands. As the World Health Organization states: 'Water is the essence of life and human dignity. Water is fundamental to poverty reduction, providing people with elements essential to their growth and development' (Brundtland and de Mello 2003). Sharing the obligation to provide safe drinking water and sound sanitation as an integral aspect of poverty reduction is the route to sustaining the health and welfare of people in the 21st century.

References

Allan, J.A. 1998. 'Virtual water: a strategic resource,' *Ground Water,* vol. 36 p. 545.

Alston, P. 1992. 'The committee on economic, social and cultural rights.' In: *The United Nations and Human Rights: A Critical Appraisal. Alston P.* (ed.), Oxford: Clarendon Press.

Ballentine, J.A. 1969. *Ballentine's Law Dictionary* 3rd ed. Rochester: The Lawyers Co-operative Publishing Company.

Beaubien, J. 2009. *Sanitation Problems Thwart Mexico's Flu Battle.* Morning Edition, April 30, National Public Radio.

Beitzel, B.J. 2007. *Biblica: The Bible Atlas.* Hauppauge NY, Barrons Educational Series, Inc.

Binnie, G.M. 1981. *Early Victorian Water Engineers.* London: Thomas Telford Ltd.

Brundtland, G.H. and de Mello, S.G. 2003. *Right to Water.* Geneva: World Health Organization.

Cane, P. and Conaghan, J. 2008. *The New Oxford Companion to Law* 1026. Oxford and New York, Oxford University Press.

Caponera, D.A. 1973. *Water Laws in Moslem Countries.* Rome: Food and Agriculture Organization of the United Nations.

Caponera, D.A. 1992. *Principles of Water Law and Administration.* Boca Raton, FL : Taylor & Francis.

Chadwick, E. 1842. *Report on the Sanitary Condition of the Labouring Population of Gt. Britain.* Edinburgh, University Press [1965].

Committee on Economic, Social and Cultural Rights 2002. *General Comment No. 15.* The Right to Water U.N. DOC. E/C.12/2002/11, 26 November.

Cullet, P. and Gowlland-Gualtieri, A. 2005. "Local communities and water investments." In Edith Brown Weiss, Laurence Boisson De Chazournes and Nathalie Berrnasconi-Osterwalder (eds), *Fresh Water and International Economic Law.* Oxford: Oxford University Press. 303–330.

Eide, A. and Alfredsson G. 1992. Introduction, in Eide, A., and G. Alfredsson, G. Melander, L.A. Rehof, and A. Rosas. *The Universal Declaration of Human Rights: A Commentary.* Oslo: Scandinavian University Press.

Frederick, K.D. 1995. *America's Water Supply: Status and Prospects for the Future*, CONSEQUENCES, *available at* http://www.gcrio.org/CONSEQUENCES/spring95/Water.html.

Glennon, R. 2002. *Water Follies: Groundwater Pumping and the Fate of America's Fresh Waters.* Washington DC: Island Press.

Hamlin, A. 1993. 'Welfare,' in Goodwin R.E. and Philip Pettit, *A Companion to Contemporary Political Philosophy.* Oxford: Blackwell Publishers.

Hirsch, A.M. 1959. 'Water legislation in the Middle East,' *8 AM. J. COMP. L.* 168, 173.

International Conference on Water and Development. Dublin Statement on Water and Sustainable Development, 31 January 1992, U.N. Doc. A/ CONF.151/PC/112.

Johannesburg Declaration on Sustainable Development, (2002). U.N. Doc. A/CONF.199/20, para. 5.

La Berge, A.F. 1992. *Mission and Method: The Early Nineteenth-Century French Public Health Movement.* New York: Cambridge University Press.

Lawrence, F 2008. 'Revealed: the massive scale of UK's water consumption,' *Guardian*, August. 20, 1.

Lauren, P.G. 2003. *The Evolution of International Human Rights: Visions Seen* Philadelphia: University of Pennsylvania Press.

Lauterpacht, H. 1968. *International Law and Human Rights.* Archon Books.

Marshall, T.H. 1950. *Citizenship and Social Class.* Cambridge: Cambridge University Press.

McCaffrey, S.C. 2005. 'The Human Right to Water,' in E.B. Weiss, L.B. De Chazournes and N. Bernasconi-Osterwaler (eds), *Fresh Water and International Economic Law,* pp. 93–115.

Mediterranean Civil Society. 2008. Statement to the Euro-Mediterranean Ministerial Conference on Water. Available at: http://www.mio-ecsde.org/filemgmt_data/files/statement%20of%20civil%20society_22_12_2008.pdf.

Melosi, M.V. 2000. *The Sanitary City.* Baltimore: John Hopkins University Press.

Nickel, J.W. 2007. *Making Sense of Human Rights* 2nd ed. Oxford: Blackwell Publishing.

Owen, D. 1964. *English Philanthropy 1660–1960.* Cambridge: Harvard University Press.

Porter, K.S. 2005. 'Should governmental water responsibilities flow downwards?' *The Journal of Water Law.* Vol 16, no. 2, 49–57.

Robertson, G. 2006. *Crimes Against Humanity* (3rd ed.). New York: The New Press.

Robinson, N.A. 1993. *Agenda 21: Earth's Action Plan*, IUCN Environmental Policy & Law Paper No. 27. Dobbs Ferry: Oceana Publications, Inc.

Rosen, G. 1993. *A History of Public Health.* Baltimore: John Hopkins University Press.

Sepulveda, M. 2003. *The Nature of the Obligations under the International Covenant on Economic, Social and Cultural Rights.* Antwerpen: Intersentia.

Tawney, Richard H. 1962, 1926. *Religion and the Rise of Capitalism*. Gloucester, Mass: Peter Smith.

UNEP. 2002. *Atlas of International Freshwater Agreements.* Nairobi: United Nations Environment Programme.

UNESCO. 2003. cited in Paul Gordon Lauren, *The Evolution of International Human Rights.* Philadelphia: University of Pennsylvania Press.

United Nations 1958. *Water for Industrial Use.* UN Report E/3058ST/ECA/50. New York: Economic and Social Council.

Webb, S. and B. Webb 1911. *The Prevention of Destitution.* London: Longmans, Green and Co.

Wells, H.G. 1940. *H.G. Wells on the Rights of Man.* Published as a Penguin special by Penguin Books Limited.

Chapter 5
Water Resources and Conflict in Lebanon

Nadim Farajalla

Background

For a considerable period of time in the middle of the 20th century Lebanon was the hub of academic, cultural, and economic activities in the Middle East. It was an oasis of democracy and tranquility in a tumultuous region and thus became known as the Switzerland of the Middle East. But as the Arab-Israeli conflict escalated and tensions rose in the Arab world, Lebanon was used as a pressure release valve and succumbed to a civil war which was fueled by external, regional powers. The combination of civil and regional strife began officially in 1975 and ended in 1990 with Lebanon in ruins (physically and economically) and a large number of its people either internally displaced or having left the country. The war saw the intervention and/or invasion of the country by Syria, Israel, the Palestinian and a multitude of western armies. Even after the official end of the internal component of the conflict, Lebanon remained subject to Israeli bombings, air raids, and incursions, the latest of which was in 2006. A major victim of the wars on and in Lebanon was its water resources and the water sector that serves the resident population. The sections that follow will attempt to contextualize and describe the impact of war on these resources and the water sector.

Brief Overview of Water Resources in the Middle East

For most countries in the Middle East, water is the limiting resource for development (Brooks, 2003). About 67 percent of the area of the Arab world receives precipitation of less than 100mm/yr, while only 18 percent receives more than 300 mm/yr (Benbiba, 2002). During recent decades, these countries have experienced significant socioeconomic development leading to an increase in water demand and an overexploitation of their water resources.

The Middle Eastern countries' main sources of water are influenced by their geographic locations. Countries in the region rely mostly on groundwater (Al-Rashed and Sherif, 2000). Countries bordering the Mediterranean, including Iraq and Jordan, rely mainly on surface water sources. Table 5.1 highlights the availability of conventional water sources in the countries forming the Economic and Social Committee of Western Asia (ESCWA).

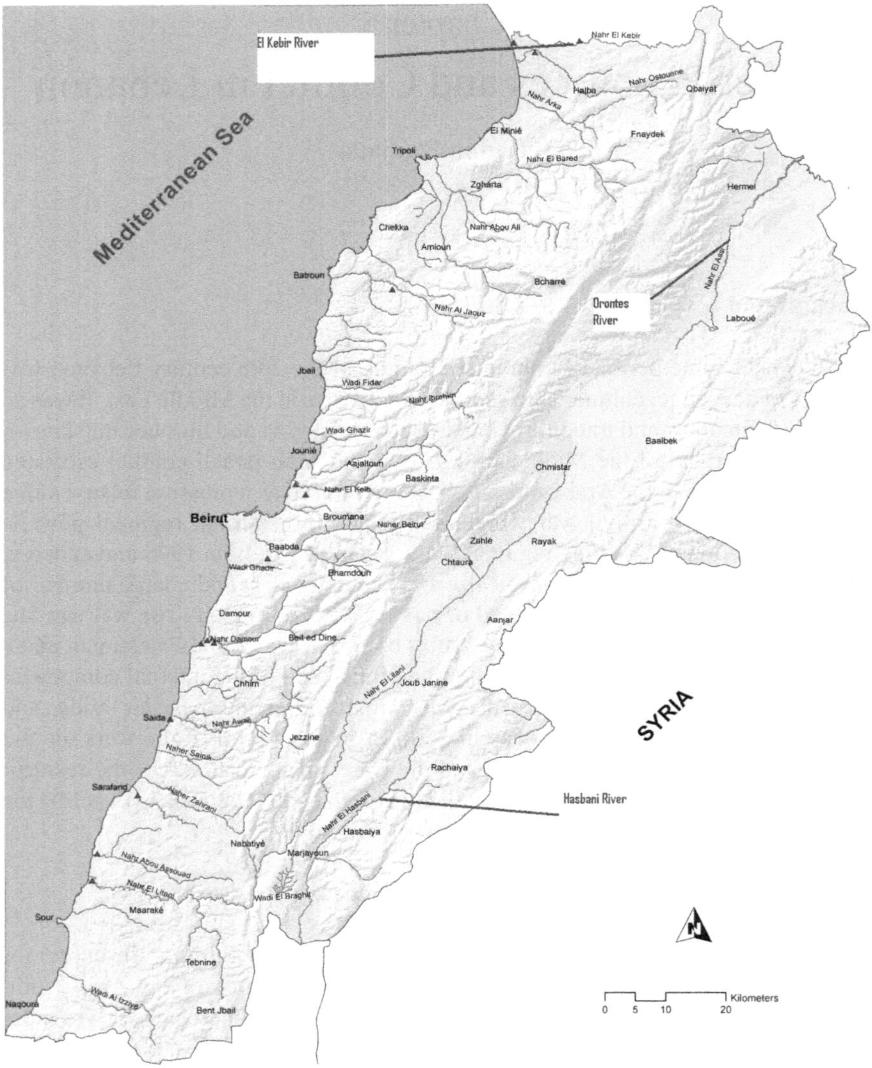

Figure 5.1 Map of Lebanon

There were three water-scarce countries in the Middle East and North Africa in 1955 (Bahrain, Jordan and Kuwait). This number had risen to 11 by 1990 (with the inclusion of Algeria, Israel and the Occupied Territories, Qatar, Saudi Arabia, Somalia, Tunisia, the United Arab Emirates and Yemen). It is expected that another seven countries will join the list by 2025 – Egypt, Ethiopia, Iran, Libya, Morocco, Oman and Syria (Darwish, 1994).

Table 5.1 Annual available water resources in ESCWA countries

	Total Renewable Conventional Water Resources (in billion cubic meters per year)		
Country	**Surface Water**	**Groundwater**	**Total**
Bahrain	0.00	0.10	0.10
Egypt	55.5	4.10	59.6
Iraq	70.4	2.00	72.4
Jordan	0.35	0.28	0.63
Kuwait	0.00	0.16	0.16
Lebanon	2.50	0.60	3.10
Oman	0.92	0.55	1.47
Palestine	0.03	0.20	0.23
Qatar	0.01	0.08	0.09
Saudi Arabia	2.23	3.85	6.08
Syrian Arab Republic	16.4	5.10	21.5
United Arab Emirates	0.18	0.13	0.31
Yemen	2.25	1.40	3.65
Totals	150.7	18.7	169.4

Source: ESCWA, 2001.

The utilization of available water resources in Arab countries has reached 69 percent and in some countries of the Arab peninsula and Persian Gulf it has exceeded 100 percent. It is estimated that by 2025, most Arab countries will fall below the internationally set critical water scarcity threshold of 500 m^3/inhabitant/year (Benbiba, 2002).

Most Middle Eastern surface water stems from three major river systems: the Tigris-Euphrates, the Nile and the Jordan. According to Berman and Wihbey (1999), the reliance of several riparians on these resources has been a catalyst for conflict, spurring confrontations such as the 1967 War (fomented by Syria's attempts to divert water from Israel) and the Iran-Iraq War (which erupted from disputes over water claims and availability). Berman and Wihbey (1999) further state that the potential for conflict arising from shared waters has spurred numerous diplomatic efforts (most notably the 1953–1955 US-brokered Johnston negotiations) and bilateral and multilateral treaty efforts, ranging from the 1959 Agreement for the Full Utilization of Nile Waters to the 1994 Israeli-Jordanian Treaty.

Water Resources in Lebanon

The approximately nine billion cubic meters (BCM) of rainfall over Lebanon leaves an impression of a Middle Eastern country rich in water resources. However, most of the 9 BCM evaporates and only 1/3 is available as a renewable resource. In fact, Lebanon could be considered a water stressed country with annual renewable water resources per capita estimated to range from 766 to 1290 m^3/capita/year. This places Lebanon well behind Iraq and Syria in terms of fresh water availability per capita (see Table 5.2).

Lebanon's water resources, though seemingly abundant, are not expected to meet the country's demands in the near future. Demand pressures from a rapidly growing population, an expanding economy, increased urbanization, and agricultural activities are leading to overexploitation and pollution of existing water resources. Ineffective management practices, fragmented institutional arrangements, lack of appropriate legislation and poor or absent allocation policies are exacerbating the situation. If the impact of climate change is added to these stress factors, then the water sources of Lebanon will indeed be insufficient to meet

Table 5.2 Renewable water per capita in selected countries in the Middle East

Country	Average Annual Precipitation (mm)	Total Renewable Water Resources (BCM)	Renewable Water Resources ($1000m^3$/capita/year)		
			1997	2015	2025
Iraq	154	63–100	2.96–4.6	1.83–2.94	1.31–2.05
Israel	630	1.5–2.6	0.28–0.44	0.19–0.36	0.14–0.31
Jordan	94	0.7–1.4	0.17–0.23	0.08–0.13	0.07–0.09
Lebanon	827	2.0–4.0	0.77–1.29	0.34–0.98	0.26–0.81
Palestinian Authority	350	0.1–0.2	0.07–0.09	0.04–0.06	0.03–0.04
Syria	252	15.0–21.5	1.16–1.44	0.76–0.95	0.54–0.61

Source: Roudi-Fahimi et al. 2002.

Table 5.3 Expected water demand in Lebanon in million cubic meters

	Domestic (MCM)	Agricultural (MCM)	Industrial (MCM)	Total (MCM)
1990	290	875	100	1263
2005	380	885	140	1405
2020	810	1310	255	2375
2025	1100	2300	450	3850

Source: Comair 2005, ESCWA 2003, World Bank 2003, Hajjar 1997.

the demands of the resident population. Projected demands on water resources per sector through 2025 are represented in Table 5.3.

Climate

Lebanon's climate is typical of the Mediterranean region and is characterized by four distinct seasons. It has a short rainy season followed by a relatively long dry period. The physiographic features of the country have a marked effect on water resources and their distribution throughout the country. The country's distinct features are its two parallel mountain ranges running north to south: the Mount Lebanon range in the west and the Anti-Lebanon in the east. The Mount Lebanon range is separated from the Mediterranean by a narrow coastal plain while the Beqa'a valley separates the Mount Lebanon from the Anti-Lebanon.

Precipitation in Lebanon is greatly influenced by the country's physiographic features. The coastal plain and the western slopes of the Mount Lebanon range receive the highest amount which is in the form of rain at lower altitudes and snow at higher ones. Next is the southern section of the Beqa'a. The plain's middle and northern portions have less rainfall and the Anti-Lebanon receives the least amount of rain. Annual precipitation on the coastal plain ranges between 600mm and 1000mm. Mount Lebanon may get precipitation up to 2000mm annually but a typical range is 1000mm to 1400mm. Rainfall in the central and northern Beqa'a is approximately 200mm to 600mm annually, while in the southern portions of the plain the rainfall is 600mm to 1000mm annually. No consistent data are available for rainfall on the Anti-Lebanon Range (Lebanese Ministry of Environment, 2001). In all regions, rainfall is concentrated in a very short period between November and April. Typically, January is the wettest month followed by December and February. The highest evaporation rate occurs in the months of July, August and September (Lebanese Ministry of Environment, 2001). Snow packs at higher altitudes may last through to May and June with an effect similar to a delay of rainfall of about three to four months. This can be seen, for example, in the comparative study of rain and the flow of Yammouneh, where flows following snow melt were delayed by about four months in comparison to those induced by rainfall (Yordanov, 1962).

Sources of Water

Lebanon is drained by 17 perennial and several seasonal rivers. Almost all of the perennial rivers are coastal with only three found in the interior of the country: Litani, Orontes (Assi), and Hasbani (see Figure 5.1). Flow from perennial and seasonal streams and rivers is estimated at around 4000 million m^3 (MCM) per year (Lebanese Ministry of Environment, 2001). Table 5.4 identifies the major perennial rivers and their direction of flow. Surface storage areas are not very abundant in Lebanon. The only major reservoir on a river is the Qaraoun Lake which is formed by the rockfill dam on the Litani River. This river which is about

Table 5.4 Mean flow rate of major perennial rivers

River	Annual Flow Rate (MCM)	Description
Major Coastal Rivers	2,470	Flow west from their source in the Mount Lebanon range: Ostuene, Aaraqa, El Bared, Abou Ali, El Jaouz, Ibrahim, El Kalb, Beirut, Damour, Awali, Saitani, El Zahrani, Abou Assouad
Kebir River	190	Flows west and traces the north border of Lebanon with Syria
Litani River	946	Drains the southern Beqa'a plain and discharges into the sea north of Tyre
Orontes (El Assi) River	400	Flows north into Syria draining the northern Beqa'a plain
Hasbani River	140	Crosses the southern border and forms one of the tributaries of the River Jordan

Source: Comair 2005, Hajjar 1997.

170 km long with a catchment of 2175 km^2, drains the Beqa'a before discharging into the Mediterranean (Comair, 2005). The river's average annual flow rate is nearly 700 MCM and the total reservoir capacity is 220 MCM (Hajjar, 1997 and Murakami and Musiake, 1994).

These resources all originate in Lebanon except for three major rivers that flow out of Lebanon or are shared with neighboring countries. These rivers are the Orontes (el Assi), which flows north into Syria, the Kebir, which traces the northern border of Lebanon with Syria, and finally, the most famous of the three, the Hasbani River, which forms the headwaters of the Jordan River. Equally famous is the Sheba'a Farms area, now under Israeli occupation, which is considered to be the recharge zone of some groundwater aquifers that formerly lay under the border of Lebanon and its two neighbors to the south and east. Of lesser fame but probably of equal importance are the groundwater aquifers that are further southwest and north of the Sheba'a region. Along the border with Israel from Ras El Naquora, on the Mediterranean coast, east to Bint Jbail and north to Tibnine are moderately deep Cretaceous karstic formations that have the potential of holding significant amounts of water. To the east of these formations are Nummulitc karstic formations that are relatively shallow and have even more abundant water supplies. Finally, there is a shared Jurassic karstic aquifer along the Mount Hermon area between Lebanon and Syria in a region that stretches from north of Marjayoun to north east of Rachaya which has significant potential (Hajjar, 1997).

Issues of Transboundary Water Resources with Neighboring States

The perceived abundance of water resources in Lebanon coupled with the severe shortage of these resources in neighboring countries has led to tensions and at times to armed conflict. As shown earlier, Lebanon shares surface and subsurface water resources with its two neighbors, Israel and Syria. With the former Lebanon shares the Hasbani/Wazzani system which forms the headwaters of the Jordan River, while with the latter Lebanon shares the Orontes and the Kebir.

Lebanon and Syria

The larger of the two rivers shared by Lebanon and Syria is the Orontes and sharing its waters has been the most contentious of Lebanon's water-related issues. The headwaters of the Orontes are in Hermel in the northern Beqa'a in Lebanon from where the river flows northward through Syria to discharge into the Mediterranean Sea. As stated by Comair (2008), the Orontes' average annual flow through Lebanon is around 400 million cubic meters. It flows for 40 km in Lebanon, 120 km in Syria, and 88 km in Turkey. In 1972 Syria and Lebanon announced that they had signed an agreement whereby Lebanon would use about 80 MCM (Soffer, 1999). Nothing of the sort materialized and the status quo of poor water management on the Lebanese side continued with only a slight portion of the allotted 80 MCM being used by residents of the northern Beqa'a. In 1976, Syria sent its army into Lebanon and became actively engaged in military activities in the country. The Syrian Army took up positions throughout the Beqa'a, but more specifically, near the headwaters of the Orontes. The Syrians thus became the de facto rulers of the area and the use of the Orontes waters was strictly controlled. After the 1989 Taif Agreement, which officially ended the war in Lebanon, Syria's military presence in the country was legitimized and Lebanon came under full Syrian control. In 1994 following brief negotiations over the Orontes waters between Lebanon and Syria, the same division was re-established whereby 80 MCM were awarded to Lebanon. However, as Comair (2008) describes, the agreement was not at all favorable for Lebanon and reflected the weak status of the Lebanese state *vis-à-vis* Syria. For example, the aquifers which fed the Orontes were not clearly identified, allowing the Syrians to extend a prohibition on drilling for groundwater well beyond the actual time-limit. Further, the waters allocated to Lebanon were those flowing during winter when the need for irrigation is minimal. This was further exacerbated by a clause prohibiting the construction in Lebanon of any reservoirs along the Orontes, thus preventing the Lebanese from using water when they needed it most – in late spring and summer. The public outcry against the agreement was such that, in 1997, an amendment to the agreement was effected which identified three main aquifers as not connected to the Orontes thus allowing the population to use their waters for irrigation. As Syria's power over Lebanon started to ebb, a further amendment was added to the agreement in 2001 which allowed Lebanon to construct a small dam on the Orontes. It also required that Lebanon construct wastewater treatment facilities in the watershed to protect the river's water quality.

In contrast to the acrimonious deliberations regarding the Orontes River, the negotiations over the waters of the Kebir River were near exemplary. Sixty percent of the catchment of the Kebir is in Syria and the rest in Lebanon. The average annual flow is about 190 MCM. In 2002, the governments of the two riparians agreed to a fair utilization of the river's waters after negotiations facilitated in part by the UN. As part of this agreement Lebanon was to have 40 percent of the annual water flow and Syria 60 percent. Further, a 70 MCM dam was to be constructed on the Kebir to regulate its flow and allow storage of winter flows.

Lebanon and Israel

Lebanon shares with Israel the Hasbani River and Wazzani Springs which have a combined average annual flow of 140 MCM. The Hasbani/Wazzani system is part of the complex of springs and rivers that form the headwaters of the Jordan River. Water has always been a contentious issue between Lebanon and Israel. Even before its inception in the late 1940s, Israel and its backers always considered water resources as a key element in ensuring Israel's viability in a water scarce region. In 1899, Theodore Hertzl advocated and Bourcart adopted a plan to include the Litani, a river whose entire flow is in Lebanon, as a major source of water for the state of Israel. Many plans were subsequently developed such as that by Lowdermilk, followed by Hays in 1948, which incorporated 400 MCM of the Litani waters to be supplied annually to Israel. The first plan to divert the Hasbani water was developed by Ionides in 1939 (Isaac and Hosh, 1992). As the Arab-Israeli conflict had just begun, such plans were perceived as a threat to all Arabs and not just to Lebanon. Thus the Arab League came up with plans of their own first advanced by McDonald and later by Bunger for the development of the Jordan headwaters including the Hasbani but excluding the Litani waters. In 1953, following severe tension over sharing of the Jordan River waters, an American negotiator, Eric Johnston, proposed a water sharing plan amongst the riparian states (Lebanon, Jordan, Syria, and Israel) which gave Lebanon only 35 MCM of the Hasbani waters (Isaac and Hosh, 1992). None of these plans came to fruition, but the Johnston Plan (published in 1955) was informally used as a basis for sharing the waters by the riparian countries of the Jordan River. However, during its occupation of southern Lebanon between 1978 and 2000, Israel strictly controlled access to the Hasbani River and the Wazzani Springs. Moreover, it prevented farmers from drilling wells for personal use so as not to influence the quantity of subsurface water flowing into northern Israel. After the liberation of southern Lebanon, the Lebanese began pumping approximately 10 MCM a year of water from the Wazzani Springs for distribution to villages that had no prior access to these waters. Even though these villages fall within the Wazzani-Hasbani basin, and the amount pumped was less than 30 percent of the amount allocated to Lebanon under the Johnston Plan, Israel threatened to bomb the pumping station. After intervention from the US, UN, and Europe and the threat from Hizbollah of retaliation for any Israeli aggression, the Lebanese were allowed to continue pumping water.

War and Water Resources in Lebanon

For nearly two decades Lebanon suffered the horrors of war on its territory; not a sector was spared, no geographic region exempt. The effects were felt on a personal level and on a national scale. The national impact of war on the water sector was often ignored, although stories of personal suffering abound. Yet the effects on the national water supply and sewerage systems led directly to the individual misery felt by thousands of Lebanese citizens. The following two sections will detail examples of damage to water resources in Lebanon due to combat activities, both on an individual level and at a national scale. The combat activities described are those experienced by Lebanon during the July 2006 War, while the personal experiences cut across the entire period of 1975 through 2006.

The War of July 2006

The July 2006 War started on 12 July 2006 and continued until 14 August 2006 when the United Nations brokered a ceasefire that went into effect on 8 September 2006 after Israel lifted its naval blockade. During the July 2006 War nearly one million people or a fourth of the country's population were displaced (UNHCR, 2006). There was extensive damage to infrastructure and public and private properties throughout Lebanon. Most sectors of the economy suffered severe losses. The initial direct damage estimate by the Lebanese government put the total damages at around US$1.144 billion distributed amongst the following (CDR, 2006):

- Transport sector (roads, airports, etc.): $484 million
- Electricity sector (production and transmission): $244 million
- Telecommunication sector: $116 million
- Water sector (well fields, conveyance networks, lift and pump stations, etc.): $80 million
- Industrial compounds (factories, warehouses, etc.): $220 million

The impact of the July 2006 war on Lebanon's surface and subsurface water resources was extensive in its geographic scope. There were direct impacts from aerial, sea and land bombardment, which led to the damage and/or destruction of structures such as reservoirs, irrigation ponds, water conveyance structures (canals, pipelines, etc.), wellheads, and pumping stations. For example, in southern

Table 5.5 Damage to Lebanon's public water infrastructure during Israel's July 2006 war on Lebanon

Regional Water Establishments (WEs)	Damage Description	Assessment (preliminary) Millions of US$
South Lebanon	• Destruction of reservoirs and water supply networks • Destruction of principle and secondary pumping and pipelines • Partial or total destruction of some pumping stations • Partial or total destruction of wastewater treatment facilities	25
Beqa'a	• Destruction of principle and secondary pumping and pipelines • Total destruction of some pumping stations • Destruction of reservoirs and water supply networks	14
Beirut and Mount Lebanon	• Destruction of reservoirs and water supply networks • Destruction of principle and secondary pumping and pipelines	6
Litani Water Office	• Damage to principle pumping stations, electric generators and equipment in the generating stations • Damage to large number of irrigation channels and networks.	12

Source: CDR 2006.

Lebanon, water supply reservoirs in Bint Jbeil, Hasbaiya, and the Wazzani area along the border with Israel were severely damaged or destroyed. In the Beqa'a, a section of Canal 900, a major irrigation canal conveying water from the Qaraoun reservoir, was destroyed, while in the southern outskirts of Beirut (Dahia), huge bomb craters, some in excess of 30 ft in diameter dotted the area exposing severed water supply and sewage pipelines. The preliminary cost of direct damages as reported by the relevant water establishments in Lebanon is shown in Table 5.5.

Indirect damage was physical and chemical. Physical damage consisted mostly of erosion and the subsequent sedimentation or deposition of eroded particles in waterbodies. Forests and orchards in many parts of southern Lebanon and the central mountain ranges were totally decimated by fires started by the various types of bombardments. These fires left the topsoil of the steep mountain slopes and valleys exposed to the erosive effects of wind and rain. What made matters worse was the presence of cluster bombs (mini-bombs delivered aerially in a large canister and designed to inflict maximum damage on groups of people). These mini-bombs and other unexploded ordinances (UXOs) prevented firefighting crews from putting out the fires and later severely restricted reforestation efforts because of the potential for serious injuries.

Indirect chemical impacts generally took the form of seepage of chemicals either from damaged facilities into groundwater resources or the transport of these chemicals along with run-off water into adjacent surface waterbodies. There was fear among residents that the water resources (surface and subsurface) might have been polluted by the bombing. Such pollution may result both directly from exploding ordinance on soil surfaces and from the spillage and discharge of chemicals from targeted sites. Spilled or discharged chemicals from targeted sites can pollute receiving water bodies and have toxic effects on soil organisms and plants at the site of spillage. These chemicals could also have detrimental effects on sites farther away through seepage into groundwater. The subterranean water can then transport these chemicals to locales farther away. Farajalla and El-Khoury (2007) reported a spike in the concentrations of nickel and chromium, both carcinogenic heavy metals, in the Ras El Ain water supply springs of the ancient coastal city of Tyre shortly after the first rains following the end of hostilities. This spike in concentration of the two metals was linked to the bombing of a region east of Tyre that formed the recharge zone of the contaminated aquifer supplying the springs.

Another such example of indirect damage is the oil spill which resulted from the bombing of the heavy fuel storage tanks serving a power generation plant in the town of Jiyyeh on the coast 20 km south of Beirut. The bombing of the Jiyyeh Power Plant on two occasions (12 and 15 July) resulted in a spill of approximately 15,000 tons of heavy fuel into the Mediterranean Sea that contaminated about 150 km of the Lebanese shoreline. The off-shore currents, moving in a northerly direction, transported the fuel all the way to the Syrian coast.

While the damage done to the water distribution and collection networks was mostly direct and short term, the damage done to the various water resources was mostly indirect and of a longer duration. Broken pipes and pumps are easily replaced or repaired, but seepage and run-off from contaminated sites takes longer to clean up. Further, as reconstruction proceeds, there is a negative impact from these activities. Reconstructing bridges and culverts invariably contaminates the water bodies in which these are constructed. Table 5.6 lists the sources of water pollution due to the combat activities.

Adaptations of Individuals to War-Induced Water Problems

The problem facing most people in Lebanon during times of open warfare was securing a safe and constant supply of potable water and water for domestic use (cleaning, washing, cooking, etc). The problem of securing a constant supply of water was particularly onerous for urban area residents. Like most urban centers in Lebanon, Beirut is crowded with buildings, with little or no room for external storage. Fortunately for residents, the water supply system in Lebanon in the pre-1975 era (and since) has been subject to rotational rationing in which one area

Table 5.6 Sources and nature of pollution of water resources in Lebanon due to the July 2006 War

Sector/Activity	Nature of pollution
Industry	
• Water Chlorination Units (Direct)	• Chlorine discharged into waterbodies
• Food Industry (Direct)	• Release of stored chemicals into waterbodies • Decomposed food stuff dumped in waterbodies reduce dissolved oxygen
• Plastic (Direct)	• Release of stored chemicals into waterbodies
• Papers (Direct)	• Dioxin, bleaching and coloring agents released into waterbodies
• Waste water treatment plants (Direct)	• Untreated waste discharged into surface waters
• Marble Industry (Direct)	• Dust and debris released into rivers
Energy	
• Electrical power generation plant (Direct)	• Fuel oil released into the sea • Potential for some of the released fuel oil to seep into coastal aquifers along with sea water intrusion
• Power generation plants – private generators (Direct)	• Diesel released into underlying aquifers
• Transformers (Direct)	• PCBs and other oils released into underlying aquifers
• Gas Stations (Direct)	• Hydrocarbons (gasoline, kerosene, etc.) leaked into underlying aquifers
Direct Military Actions	
• Explosions, bombings, and fires (Direct)	• Nitrogen (and potentially phosphorous) discharged into waterbodies • Soot, sediment, dust reaching surface waterbodies
• Movement of military vehicles Truck (Direct)	• Hydrocarbons leaked into waterbodies • Geomorphologic damages – demolition of stream banks and blockages of streams and wadis
• Unexploded ordnance (Indirect)	• Restricting access to affected areas

Table 5.6 Continued

Construction	
• Debris and trash from residential buildings	• Household waste/chemicals, sediment, etc. are dumped into the sea, rivers, and other open waterbodies • Household waste/chemicals, sediment, etc. are dumped into pits which would later leak into the underlying aquifers
• Debris and trash from industrial plants	• Production waste/chemicals, sediment, etc. are dumped into the sea, rivers, and other open waterbodies • Production waste/chemicals, sediment, etc. are dumped into pits which would later leak into the underlying aquifers
• Bridges	• Intrusion into the stream • Destruction of stream banks and river beds • Blockage of streams by temporary crossing points during bridge construction phase (some with no provision for stream flow) • Sediment discharge into streams • Hydrocarbon leaks from construction vehicles and from vehicles using temporary crossings
Agriculture	
• Dairy and meat (cattle, mutton, goats, etc.) • Poultry • Fisheries	• Improper disposal of rotting carcasses in rivers and ponds leading to high BOD and low dissolved oxygen in the water • Disposal or accidental dumping of feed into waterbodies • Disposal or accidental dumping of veterinary medicines into waterbodies • Nuisance odor of decomposing carcasses and feed
• Destruction of stored fertilizers, pesticides, stored chemicals	• Discharge or leakage of nutrients and other chemicals into water bodies
• Destruction of canals	• Waterlogging at site of destruction
• Destruction of greenhouses and other equipment	• Dumping into rivers and open waterbodies
Transport	
• Roads	• Same as Bridges in “Construction” item

Source: Farajalla 2007.

of a city would get water supply on a particular day at a specific time for a given duration and an adjacent area would get water for a similar duration. This led to reservoirs being installed at the top (or bottom) of nearly all buildings to enable a continuous water supply for their residents. Nevertheless, such a feature did not suffice during times of extended shortages and the situation was made worse by the frequent power outages which prevented the pumping of water to roof-top storage facilities. Thus people started filling up bathtubs and other large sized containers with water whenever it flowed through the tap. As the water supply dropped from near 250 liters/person/day (l/p/d) to near 25 l/p/d (Acra, S., 2006) and government authority weakened, people started drilling private wells. This activity became so widespread that seawater intruded into many coastal aquifers rendering them useless (Acra, A., 1992) until the time that sufficient recharge has taken place.

The quality of the stored and pumped water became more questionable as the years progressed and the sale of bottled water grew rapidly. Not many urban dwellers could afford the luxury of buying bottled water and they had to resort to boiling available water. Since necessity is the mother of invention, a Lebanese scientist was able to develop a method of disinfecting available water to levels that were acceptable by UNICEF and the World Health Organization (Acra et al. 1984). Aftim Acra, a renowned professor of public health at the American University of Beirut developed this simple procedure, by which water is placed in clear plastic or glass bottles which are then exposed to sunlight for about an hour or so. Ultraviolet rays then killed microorganisms in the containers rendering the water safe for drinking.

Conclusion

For nearly two decades Lebanon, its people, and its resources suffered from the horrors of war. The weakened state of the nation during these decades and since has exposed its water resources to the greed of its more powerful neighbors. Further, the country's water resources were affected directly through pollution and over exploitation and indirectly through official neglect and mismanagement. Many Lebanese have lived with the credo of 'no state' in which each individual is left to fend for her or himself with no regard for the general good. If this situation continues and if the desires of its neighbors to acquire Lebanon's water are not satisfied, not only will the country's water resources be permanently lost; the entire country will cease to exist.

References

Acra, A. 1992. Sensitive criteria for the diagnosis of seawater infiltration into groundwater. In *UNICEF: Proceedings of the First Symposium on Water in Lebanon*. Beirut.

Acra, A., Raffoul Z. and Karahagopian Y. 1984. *Solar Disinfection of Drinking Water and Oral Rehydration Solutions: Guidelines for Household Application in Developing Countries.* Published for UNICEF by Illustrated Publications, SAL, Beirut, Lebanon.

Acra, S. 2006. Impact of war on the household environment and domestic activities: vital lessons from the civil war in Lebanon. *Journal of Public Health Policy*, 27 (2), 136–145.

Al-Rashed, M.F., and Sherif M.M. 2000. Water Resources in the GCC Countries: An Overview. *Water Resources Management*, 14 (1), 59–75.

Benbiba, M. 2002. Integrated Development of Wadi Systems, in: Whater H. and Al-Weshah R. (ed.), *Hydrology of Wadi Systems-IHP regional network of Wadi hydrology in the Arab region, in cooperation with the Arab League Educational Cultural and Scientific Organization (ALESCO) and the Arab Center for Studies of Arid Zones, IHP-V Technical Documents in Hydrology* No.55. Paris: UNESCO, 113–121.

Berman, I and P.M. Wihbey. 1999. The New Water Politics of the Middle East. *Strategic Review*, 27, 45–52.

Brooks, D. 2003. Between the great rivers: Water in the Heart of the Middle East. In: Rached, E., Rathgeber, E. and Brooks, D. (eds) *Water Management in Africa and the Middle East: Challenges and Opportunities*. Ottawa, Canada, International Development Research Centre (IDRC), pp. 1–17.

Comair, F.G. 2005. *Lebanon's Water Resources – Between Waste and Investment* (in Arabic) Beirut.

Comair, F.G. 2008. Gestion et hydrodiplomatie de l'eau au Proche-Orient. *L'Orient-Le Jour*, Beirut, Lebanon.

Council for Development and Reconstruction (CDR). 2006. *Preliminary Estimate of the Damage Caused by July 2006 War*. Government of Lebanon. Beirut, Lebanon.

Darwish, A. 1994. Water wars. *Lecture given at the Geneva Conference on Environment and Quality of Life*, Geneva.

ESCWA. 2003. *Updating the Assessment of Water Resources in ESCWA Members countries*, United Nations.

ESCWA. 2001. *Development of freshwater resources in the rural areas of the ESCWA region using non-conventional techniques*. United Nations, New York, 1–81.

Farajalla, N.S. 2007. War and water resources. In*:* Khoury, R. (ed.) *Rapid Environmental Assessment for Greening Recovery, Reconstruction and Reform – 2006*. Lebanon, UNDP.

Farajalla, N.S. and El-Khoury, J. 2007. Impact of the July 2006 conflict on the water quality at the Tyre coast Nature Reserve – a Ramsar site in Lebanon. *Wetlands*, 27 (4), 1160–1164.

Haddadin, M.J. 2005. *The Jordan River Basin – Water Conflict and Negotiated Resolution*. UNESCO.

Hajjar, Z. 1997. Lebanese Water and Peace in the Middle East. *Dar al Ilm Lil Malayeen*, Beirut, Lebanon (in Arabic).

Lebanese Ministry of Agriculture. 2003. *National Action Programme to Combat Desertification*. Lebanon: Ministry of Agriculture.

Ministry of Environment. 2001. *State of the Environment Report*, Beirut: Ministry of Environment.

Murakami, M. and Musiake, K. 1994. *The Jordan River and the Litani, in International Waters of the Middle East: From the Tigris-Euphrates to the Nile.* Oxford: Oxford University Press, 128–129.

Roudi-Fahimi, F., Creel, L. and De Souza, R.M. 2002. *Finding the Balance: Population and Water Scarcity in the Middle East and North Africa*. MENA Policy Brief, Population Reference Bureau, Washington DC.

Soffer, A. 1999. *Rivers of Fire: The Conflict over Water in the Middle East.* Plymouth: Rowman & Littlefield.

United Nations Office for the Coordination of Humanitarian Affairs (UNHCR). 2006. *Situation Report 39 – Lebanon Response – 20–27 September 2006*: http://www.reliefweb.int.

World Bank. 2003. (DRAFT). *Irrigation Sector Sustainability Policy Note.*

Yordanov, Y.V. 1962. *Aperçu succinct sur l'hydrogéologie du Liban.* Dar El Fann, Beyrouth.

PART III
Managing a Scarce Resource

Chapter 6
Options for Sustainable Water Management in Arid and Semi-Arid Areas

Tammo Steenhuis
With contributions from Gil Levine, A. Volkan Bilgili, Eloise Kendy, Zeynep Zaimoglu, M. Ekrem Cakmak, Mehmet Ali Cullu and Sahin Ergezer

Competition for water made scarce by intensive irrigation and unequal distribution is already a major source of conflict in the arid and semi-arid Middle East, which is the region of the eastern Mediterranean most commonly associated with potential crisis. The stresses on water quantity and quality in the arid and semi-arid countries that border the eastern Mediterranean are, in fact, quite varied. To take just two examples, tourism, which appeared to offer an endless source of income to the rocky, arid islands of the Mediterranean, has become an environmental and agricultural liability, particularly in places like Malta and the Aegean islands, which now import almost all their water; Turkey's GAP project, designed to aid development in the backward Southeastern Anatolian region through the construction of large dams, has created a potential for conflict with the countries below the dams, namely Syria and Iraq. What is common to the region is a concern with the insufficient supply or deteriorating quality of fresh water, and the potential this situation creates for human suffering, political conflict, and environmental degradation.

Insufficient supply or deteriorating quality of fresh water is a result of the pressure put on our natural resources by large increases in population. Humans have always used the natural resource base for feeding themselves. We can distinguish roughly five stages in the human development of the natural resource base. Initially, when the pressure on the natural resource base is low, food is gathered from land or sea. Cultivation of crops does not take place, and the water needs are minimal. In a second stage, when the forest or sea does not supply sufficient food for an expanding population, agriculture begins and crops are cultivated using the available rainwater in locations that are sufficiently wet during the year. At this stage, the food production per unit area is not relevant because the area can be infinitely extended until there is sufficient production. When more land must be developed to feed an ever-increasing population, demands for more food have to be met by increasing the yield. This is achieved by using better management practices, including adding fertilizers but also preventing erosion so that the valuable top soil will not wash away. In the fourth stage, increasing yield without adding additional water becomes unfeasible and societies begin investing in more advanced irrigation technology. The water used originates either from rivers, reservoirs or groundwater. This is expensive and therefore a last resort. At this stage wetlands are also cultivated, where the risk of flooding is high. Finally, in the fifth stage when all available water is in use, and what has not been used is either too salty or too toxic for irrigation, saltwater management becomes the dominant factor.

The watersheds around the Mediterranean are currently in stages 4 and 5 with the exception, perhaps, of the east coast of the Adriatic Sea. Competition for the available water among urban, industrial and agricultural users is fierce. For planning purposes as well as the avoidance of conflict, understanding the total quantity of available water in each basin and the effects of water-saving measures is important. In many cases, the water savings that are claimed are in effect not water savings at all. A good example is in irrigation systems, where water savings are claimed through switching from inefficient flood irrigation to more efficient sprinkler irrigation. Similarly, in the Nile Valley in Egypt it is sometimes stated that the water that is applied in excess of evaporation is 'lost' and more efficient irrigation would make more water available. This is a false claim, since the 'lost' water recharges the underlying shallow groundwater table which in turn will rise and the water will flow back to the Nile. Thus, the 'lost' water is reused downstream until it becomes too salty (stage 5). Of course when irrigation takes place in the Sahara desert, the leached water will end up in the deep groundwater and can be considered lost unless it is pumped up (or appears in one of the desert oases where the regional groundwater surfaces).

To find out how much water is available in hydrological terms, we must look at the water balance in a watershed. A watershed is all the land that drains to a defined point or 'outlet'. Moving the outlet downstream will therefore increase the watershed. There are two water balances for a watershed: one for the surface and vadose zone, (which includes all the surface waters and the unsaturated soil) and one for the aquifers. In calculating a water balance, only terms that specify fluxes across the boundaries of the watershed are included. Internal fluxes are important for what is done with the water, but as long as they do not affect the transport over the watershed boundaries, they are not considered in a calculation of the water balance of the watershed. For the surface (and vadose) balance the calculation is as follows: the precipitation P, evapotranspiration by the plants and soil ET, irrigation from groundwater I, the recharge to the groundwater R and water flowing in the river past the outlet Q. The surface (and vadose) water balance can be written in its most simple formula as:

$$P + I - ET - R - Q = \Delta S \qquad (1)$$

where ΔS_r is the change in the amount of water stored in the watershed. For example, if it rains it takes some time before the water reaches the outlet and we need a term to keep the balance. All units in Eq. (1) are in depth (cm) or a volume per unit area. The water balance tells us that over a sufficiently long time all rainfall is either evaporated (and is lost to the atmosphere), flows out to the sea, or increases the groundwater level. The groundwater reservoir balance is simpler and can be written as:

$$R - W = \Delta S_g \quad (2)$$

where W is the water that is either released naturally as base flow and/or withdrawn by pumping from the aquifer and applied to the crops, and ΔS_g is the change in groundwater storage. When ΔS_g is divided by the drainable porosity, it equals the change in height of the groundwater table. However, to pump water out of the aquifers it has to be conductive and consist of sand and/or gravel or sandstone.

Water can also be stored temporarily. When the watersheds have shallow soil or are sloping there is no way to store the water unless reservoir dams are built. When the soil is deep, rain can be temporarily stored in the groundwater. The groundwater can be used for irrigation, drinking water, or industrial purposes by pumping it to the surface and can thus be depleted. Storage in the form of groundwater or in reservoirs is extremely important in arid and semi arid areas and ideally can be used to store water from wet years to bridge the water shortage of drought years. However, this ideal is often not met and groundwater in these areas is being steadily depleted. An example from the North China Plain indicates what politically unacceptable sacrifices have to be made not to draw down the groundwater table.

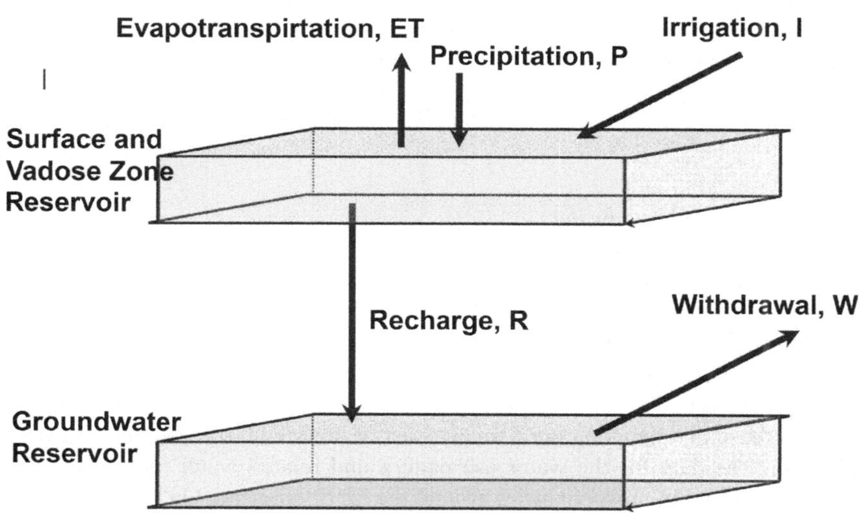

Figure 6.1 Water cycle for the subsoil: surface and (vadose zone) and groundwater reservoir

Water quality is as important as quantity. River water contains small quantities of salt originating either from rain or from being picked up on the way to the river. Unlike water, salt is conservative. When water evaporates, salt stays behind. Decreasing flows will increase the salt concentration. There is no hydrologic upper limit for the salt that can be stored in the watershed. However, when too much of the water evaporates as it travels down the watershed to the outlet and is used repeatedly, it becomes toxic to plants and humans. So in stage 5 the salt balance determines what fraction of the rainwater can be used before it becomes unsuitable for human consumption.

We will first look at the water balance for regions that are irrigated with groundwater. Although our example is the North China Plain, the findings apply equally to the Mediterranean basin where this type of irrigation has already depleted many aquifers and in coastal areas sea water has replaced the once pristine groundwater used for potable water. In addition, we discuss the water balance of the entire Nile Basin and how it will change when dams are built. We also address the salinity issue in regions that are irrigated from reservoirs and establish trends in salinity concentration in these regions and what measures can be taken to prevent salt build-up. The newly-built dams in the Anatolia region of Turkey offer an example where the salt build-up has made large areas unsuitable for crop production.

Case Studies

The three case studies discussed are all located in rainfall-deficient areas. They encompass various forms of irrigation: from shallow aquifers water underlying the area; from reservoirs with dams impounding water from watershed upstream; and from the river directly in case of the Nile. In the analysis the same water budget method is used but the management options vary greatly between the case studies.

Irrigation from Groundwater

In rainfall deficit areas the shortage between rainfall and crop requirements can be met by irrigation from aquifers underlying the plain. Pumping water from the aquifers has led to continued decline of groundwater levels despite improved irrigation efficiency and reduced pumping. Although this has been shown to be true for the North China Plain, this apparent contradiction is true for many aquifers such as those that lie along the African coast of the Mediterranean Sea. However, obtaining the data for the water use pattern and related water table declines is time-consuming in most countries around the Mediterranean. On the other hand a study of the water balance in the North China Plain where the data is available and various drastic water management practices have been used in the last 50 years, is ideal for showing what management practices are needed to stop mining the groundwater table.

In the North China Plain Agricultural policies and related water resource development policies have undergone four distinct phases. Before 1949 there was no major irrigation development and a single rain-fed crop was grown. Periodic flooding occurred and poor drainage was a major problem at that time. Irrigation development began during the Nation Rebuilding phase (1949–1958). In the Commune Era between 1958–1978, groundwater irrigation began in earnest and led to improved crop yields and continuous cropping with two harvests each year. Even at this early stage, declines in the water table were evident (Fig 6.2). Stream flows into Luancheng also decreased. In the Early Reform period (1979–1984) production was de-collectivized and groundwater pumping for irrigation decreased from about 1,020 mm/year in 1976 to about 390 mm/year in 1996. Nevertheless, water table declines continued and concerned regional authorities formulated regulations to strengthen groundwater management.

Thus, ironically, the development of irrigation systems left the county with less water than before irrigation began and despite the great changes in the amount of irrigation related to the different policies, the groundwater decline was steady after irrigation began. To explain these apparent contradictions in the water balance, the approach introduced earlier is followed (Eqs. 1 and 2).

In Luancheng County the average annual precipitation (P) for the study period was 46 cm; the average evapotranspiration (ET) from a double cropped system calculated with the Thornthwaite Mather Procedure was 66 cm (Kendy et al., 2003) and for a single cropped system without irrigation it was 33 cm plus an additional

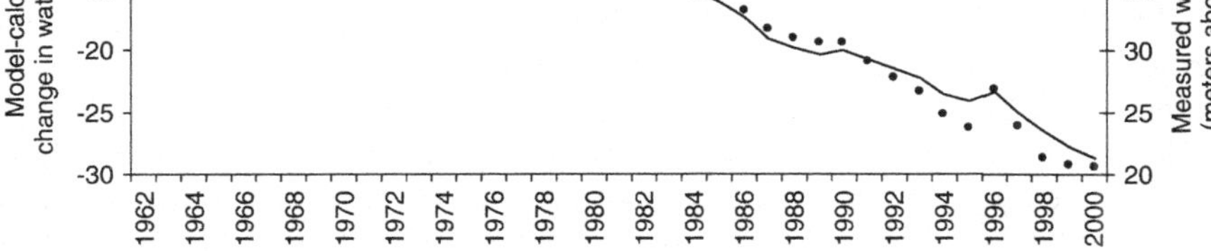

Figure 6.2 Hydrograph showing historical water-table elevations beneath Luancheng Agro-Ecological Research Station, Luancheng County, Hebei Province 1974–2000. The solid line is the observed water table. The symbols are the predicted groundwater table elevations (Kendy et al., 2003)

10 cm evaporation from the bare soil during the rainless period. Substituting these values into Eq.1 we obtain an equation for the root zone for a single crop without irrigation:

$$R = 46 - 43 = 3 \quad \text{(cm)} \tag{3}$$

and using Eqs. 2 and 3 in the groundwater reservoir balance equation with withdrawal, W, set to zero we obtain the following:

$$\Delta S_g = +3 \ \text{(cm)} \tag{4}$$

Thus, on average there is a net recharge with a single crop and the groundwater will rise until it discharges in the form of springs. The main city in Luancheng County, Shijazhuang, was also known as the city of the springs, anecdotal confirmation of the water balance. Since the 'ΔS's' in equations 1 and 2 are zero over the long term, the flow at the outlet Q = 3 cm. This suggests that there is a depth of 3 cm of water lost over the whole watershed per year and the volume is equivalent to the area of the watershed multiplied by 3 cm.

Now if we determine the water balance for the double cropped system we obtain for the root zone

$$R = 20 - I \tag{5}$$

and for the groundwater when all the irrigation water is withdrawn from the aquifer and hence W=I, Eq. 2 :

$$\Delta S_g = -20 \ \text{(cm)} \tag{6}$$

Thus, the decrease of storage is 20 cm per year and is equal to an annual drop in the water table of 1 m (drainable porosity 20 percent) as observed in Figure 6.2, since there is no water to fill the stream Q = 0. Unexpectedly at first, the recharge and the withdrawal term drop out of the groundwater balance equation (Eq. 6) and the decline in the groundwater only depends on the difference between the evaporation and the rainfall at the surface. Conceptually it is intuitive since the lateral fluxes are small when which we have a situation similar to a (large) cylinder where we circulate water from the bottom to the top with a net loss at the top of the cylinder. In cases where deep groundwater is used such as in Sahara desert in Libya and then transported to the coast, the recharge ends up in the aquifers near the coast, recharging these depleted aquifers. All water pumped from the Sahara desert is gone.

In practical terms this means, for the Mediterranean Basin, that in all cases where irrigation is practiced using the shallow groundwater and the crop uses more water than the rainfall, the water table will decline. The amount of water that a crop uses when irrigated is approximately equal to the days that the crop(s) is in the field times the daily potential evaporation rate of approximately 5 mm/day. If the rainfall is less, groundwater use is probably not sustainable. This is valid for most crops with the exception of olive tree plantations which can produce a crop with much less than 5 mm/day. Thus, for the same piece of land, an olive tree plantation could be sustainable while kiwi trees are not.

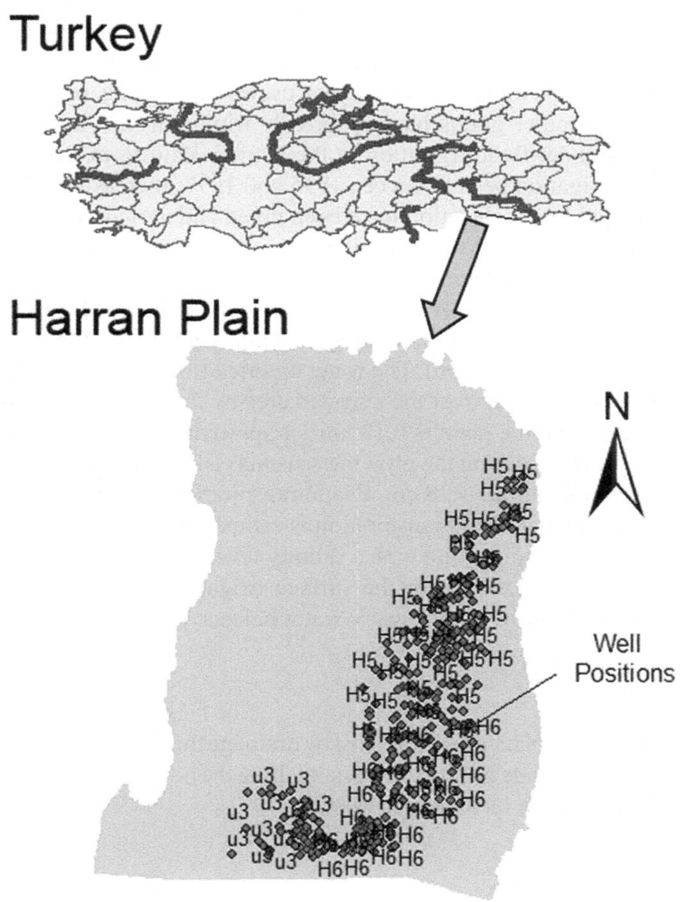

Figure 6.3 Harran Plain in Turkey. The closed circles are the location of the observation wells. The letter number combinations are the well indicators used in the text (Steenhuis et al. 2006)

Irrigation from Reservoirs (Harran Plain, Turkey)

Irrigation of the Harran Plain (Figure 6.3) is part of the Southeastern Anatolia Project (GAP), a massive $32 billion public project to harness the hydroelectric power of the upper reaches of the Tigris and Euphrates rivers and to irrigate the fertile plains that lie between them. The Harran Plain is irrigated with water from the Ataturk Dam, the sixth largest rock-filled dam in the world. It was completed in 1993, and currently generates 8.9 billion kWh of electricity (Anonymous, 2006).

The recently introduced irrigation has increased agricultural activity from one crop per year to five crops over a two-year cycle. Crop yields of cotton, wheat, barley, lentils and other grains have reportedly tripled in the Harran plain. The amount of land under flood irrigation in the Harran plain from the Ataturk dam is 121,000 ha. Irrigation began between 1995 and 1997. Increasing salinity levels in wells located at lower elevation have been noted and agricultural land is being taken out of production due to high salinity levels (Kendirli et al., 2005). This is depicted in Figure 6.4c, where salt concentration in 3.5 m deep wells is shown as a function of their elevation for the years 2003 and 2004. In order to understand the best management options for reducing the salinity problems, the water balance is calculated by Steenhuis (2006). The water diverted from the dam for irrigation is 1.98 km^3 or 99 cm per year over the irrigated area by flood irrigation. The amount of water drained from the plain is 0.19 km^3 or approximately 7 cm over the flood irrigated area. Rainfall during the growing season is a little less than 2 cm per year. The average annual rainfall is 28 cm. Evapotranspiration over the growing season is 82 cm and the annual evapotranspiration is estimated at 100 cm. The quality of the irrigation water is quite good with a salinity level of 0.6 mmhos/cm.

Using equation 1, we find for the surface irrigated area that the net excess water (or recharge) for the whole year's water balance in the root zone is:

$$R = 99 + 28 - 100 = 27 \text{ (cm)} \tag{7}$$

For the groundwater balance the water lost by drainage through natural groundwater outflow equals an average of 7 cm. For establishing the net gain in the groundwater, we calculate net storage increase per year as:

$$\Delta S_g = 27 - 7 = 20 \text{ (cm)} \tag{8}$$

In addition to the water balance there is also a salt balance for the irrigated area. Since salts do not volatilize, the mass balance is simply what is being applied to the field by irrigation water minus what is lost in drainage water. The irrigation water delivered to the Harran Plain has an electric conductivity of 0.6 mmhos/cm, which is equivalent to a salt concentration of 380 mg/l (Richards, 1954). Assuming that drainage and irrigation water have the same salt concentration, then 99 cm of irrigation water minus the 7 cm of drainage represents approximately 3,760 - 260 = 3,500 kg of salt that is added on the average each year to one ha of land.

The wells installed were only 3.5 m deep and therefore cannot indicate the permanent water table depth when it is below the bottom of the wells. Any water in the well is from temporarily perched water tables. The concentration pattern depicted in Fig. 6.4c is consistent with a rising groundwater table as we will show below. Note that as soon as the groundwater is less than 2.5 m from the soil surface, water begins evaporating (Mehani, 1988). However, what is even more important is that a shallow groundwater table prevents leaching of the salts that are carried with the irrigation water. These salts remain in the root zone, increasing the salinity.

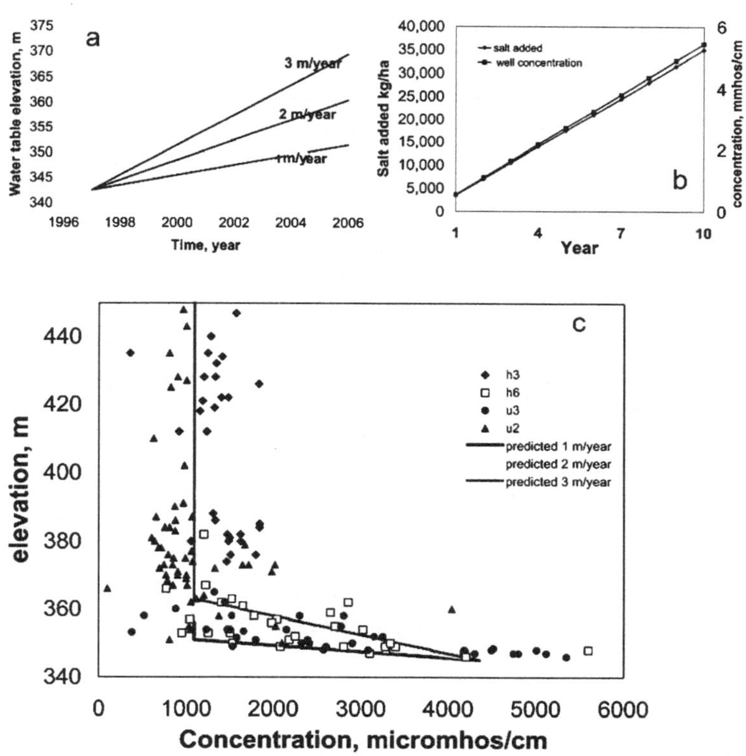

Figure 6.4 Prediction of well water salinity concentrations. a: water table height as a function of the yearly rise of 1, 2 or 3 m/year; b: concentration in wells as a function of the years that the groundwater is within 2.5 m from the surface and c: predicted (lines) and observed (points) concentrations for the wells located in the Harran Plain. The well identifiers are the same as used in Figure 6.3

Making the reasonable assumption for a semi arid climate without irrigation that the water table is horizontal, the wells at the lowest elevations (345 m in Fig 6.4c) are affected first by a rising groundwater table. Thus, these wells at the lowest elevation accumulate salts longer than wells at higher elevations. Therefore, for a rising water table, wells at lower elevation will display the highest salinity levels and the concentration will decrease with increasing elevation of the wells. For the wells with elevation above 360 m in Fig. 6.4c where the groundwater table is below the bottom of the wells, the salt concentration of the water in the root zone depends on the amount of leaching that takes place. When much more water is added than needed by the plants, the concentration approaches that of the irrigation water. For smaller leaching fractions, concentrations can be several times that of the irrigation water.

Since we do not have data from before 2003, we do not know when the groundwater rose near the surface, but based on the variation in the data depicted in Figs 6.4a and 6.4c we can estimate when the concentration began to increase rapidly. To approximate the rise of the groundwater, we calculate the concentration in the well water for a rising groundwater of 1, 2 and 3 m per year (Fig 6.4a) and compared this to the observed values in Figure 6.4c. As soon as the water table is within 2.5 m, the salts added in the irrigation water will not leach and will accumulate near the surface. In Figure 6.4b the amount of salts accumulated near the surface is given as a function of the number of years that the groundwater is within 2.5 m from the surface. By combining the results in Figs 6.4a and 6.4b, we obtain the predictions of well concentrations in the saturated zone in Fig. 6.4c. In the unsaturated zone we assumed that the concentration was twice that of the irrigation water because the rainfall and excess irrigation amounts were approximately half of the amount irrigated. Despite all the assumptions made to calculate the concentration in Fig 6.4c, the 1 m/year water table rise brackets the lower envelope of observed concentration in the groundwater and the 3 m/year, the upper envelope of the observed concentration in the wells. This is in reasonable agreement with the water balance calculations, since the net recharge to the groundwater of 20 cm/year (Eq. 8) in a clay soil with a drainable porosity of 10 percent results in an average increase in water table depth of 2 m/year.

Egypt and the Water from the Nile

The final example is the water supply for Egypt. As discussed elsewhere in this book, the only water available in Egypt (ignoring desalinization and some groundwater from the Sahara desert) comes from rain that falls in the Nile watershed that stretches from Rwanda, Kenya, Congo, to Ethiopia and even Eritrea. 85 percent of all water that reaches Egypt comes from Ethiopia and anything that Ethiopia does with the water makes Egypt nervous. As we will see, this cannot be justified technically. In order to see the effect on the water supply to Egypt we will examine the Karadobi dam on the Blue Nile in Ethiopia. A feasibility study is underway for this dam (Awulachew et al., 2009). It will be a 250 m high concrete dam with a reservoir volume of 40 km^3 approximately half of which is storage that can be used for hydropower. The reservoir area is 460 km^2. (*http://en.wikipedia.org/wiki/Lake_Nasser*) The water balance is more complicated than before because of the monsoon where there is excess water that runs off. This water runs off with or without a reservoir at the particular location. Thus, during the months with excess precipitation, the reservoir and plants evaporate the same amount and there is no water lost during the wet months by installing a reservoir. Consequently, when building a reservoir the only extra evaporation that occurs is during the rainless period which lasts approximately seven months for the location where the reservoir will be built on the Nile. The average evaporation is 4 mm/day in the Ethiopian highlands. Thus we can calculate the volume of water that is lost to evaporation by building a reservoir; this is the product of the surface area of the reservoir times the number of days that there is little precipitation:

$$460{,}000{,}000\ m^2 \times 210\ \text{days} * 0.004\ \text{m/day} = 400{,}000{,}000\ m^3 \text{ or } 0.4\ \text{BCM}$$

If we compare this to Lake Nasser in Egypt which has a storage capacity of 170 km^3 with a live volume of 132 km^3 and a surface area of 5,250 km^2 with little precipitation and an evaporation rate of 6 mm/day, we get an annual loss of evaporation of approximately 10 BCM. Thus five dams like the Karadobi dam will give more total storage than Lake Nasser with an evaporation loss of 2 BCM. In order to meet the same 'live' storage amount as Lake Nasser, seven Karadobi dams are needed with an evaporation loss of a little less than 3 BCM.

Options for the Management of Water

Egypt is in the last stage (5) in the development of the natural resource base. Only water that is excessively salty or polluted flows from the Nile into the Mediterranean. Thus, from the mass balance perspective (Equation 1) the only way that Egypt can increase its long term water supply is by having less evaporation in the Nile Basin, since it is unlikely that precipitation can be increased. One of the popular options for Egypt to have less evaporation is by building the Jonglei diversion around the Sudd swamp in Sudan. It will decrease the size of the swamp and thereby decrease the total amount of water. It is estimated that the Jonglei diversion project would produce 4.8 BCM of water per year (*http://en.wikipedia.org/wiki/Sudd*). There are, however, complex environmental and social issues involved, which may limit the scope of the project in practical terms.

Similar or greater savings in water by reducing evaporation to those predicted for the Jonglei diversion could be achieved by storing the water in Ethiopia instead of Lake Nasser (see above). An additional benefit of storing water in Ethiopia is that there is more 'reservoir space' and water that was spilled or deliberately leaked in the past during the wet years—such as in 1998, when the water was leaked westwards into the Sahara Desert—can be stored for future use. The spilled water formed the Toshka Lakes (*http://en.wikipedia.org/wiki/Lake_Nasser#cite_ref-asdf_0-0*). Politically, of course, building dams in Ethiopia is not without consequences for Egypt because once the water is stored, it can be used for irrigation in Ethiopia which then increases the evaporation. There is also a possibility that by increasing the water level through the construction of dams in Ethiopia, water can be pumped out of the basin so that none of the Blue Nile water will end up in Egypt.

Unlike reservoirs, where the water evaporates, storage as groundwater has the advantage that evaporation losses are minimized (at least as long as the groundwater stays below a depth of 3–5 m). For that reason the Toshka lakes could be seen as the groundwater reservoir and thus a water-saving source for future generations. However, from the limited information available, the decline in the lake's water levels is approximately equal to the potential evaporation, making recharge of groundwater small (El Bastawesy et al., 2007).

Managing groundwater in stages 4 and 5 is not simple. To halt declining groundwater levels, an increase in the efficiency of irrigation is generally advocated. However, as was demonstrated for the North China Plain, the only way to stop the decline in water table levels is to keep precipitation (and irrigation with water other than groundwater) above the evapotranspiration level. In agricultural areas, if no other water sources are available, adopting less water-intensive cropping patterns is needed. In Luancheng County in the North China Plain, water balance studies by Kendy et al. (2004) have shown that any cropping routine which includes a second crop (winter wheat) cycle will not show a significant reduction in groundwater depletion (Fig. 6.5). Evaporation from one crop is approximately equal to the rainfall in Luancheng County. A second crop will put the evaporation over the rainfall and

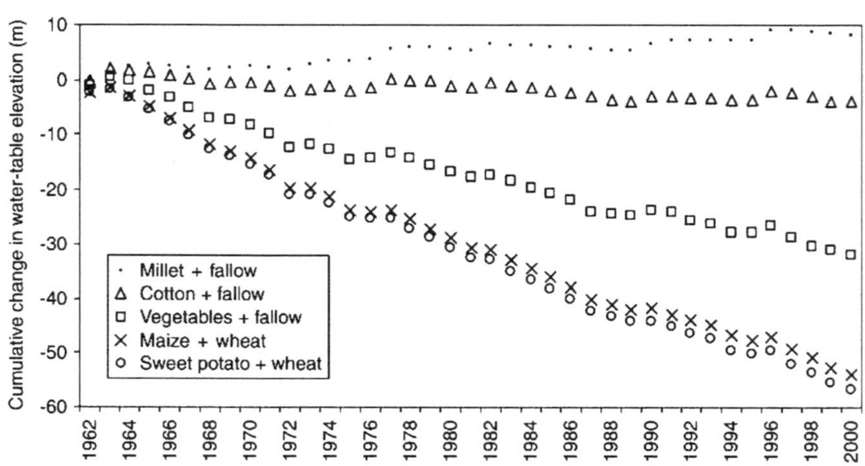

Figure 6.5 Predicted groundwater levels for different cropping patterns in the North China Plain

the extra irrigation that is needed will cause a decline in the groundwater table. These simple calculations vary depending on the amount of rainfall in the Mediterranean Basin but the conclusion is equally valid that the only way to decrease water table decline (when not growing olive trees) is by reducing the number of crops or importing water from other basins . Unfortunately, growing fewer crops is not an option that is socially and economically palatable to most people. It may, in the future, become an option if sufficient alternative employment is available besides agriculture. In this case, food can be imported and, for example, the water equivalent of 1 m^3 for each kg of imported grain would be saved. Another option is the transformation of land from rural to urban. While specific data is not available, it is commonly accepted that urban land use depletes much less water than crop production but at the same time might pollute it with toxic chemicals.

More intensive water use in stage 4, when irrigation from a reservoir is involved, leads to increased salinization. In arid and semi arid areas, this is a worldwide phenomenon that was demonstrated earlier with the example of the Harran Plain. Increasing irrigation efficiencies will slow down the rise in the water table but not arrest it, since extra water always needs to be added to leach the salt.

The only option that remains, then, is building a drainage system. Drainage systems are effective both because the groundwater table is kept at a depth below which there is little evaporation and because excess salts can be flushed from the root zone. The cost of installing a drainage system is high (about the same order of magnitude as the cost of irrigation).

Another possibility is pumping the groundwater so that the water table will remain at the same depth. Irrigation will still be needed to maintain the new water table height, because otherwise the water table will decline as in the North

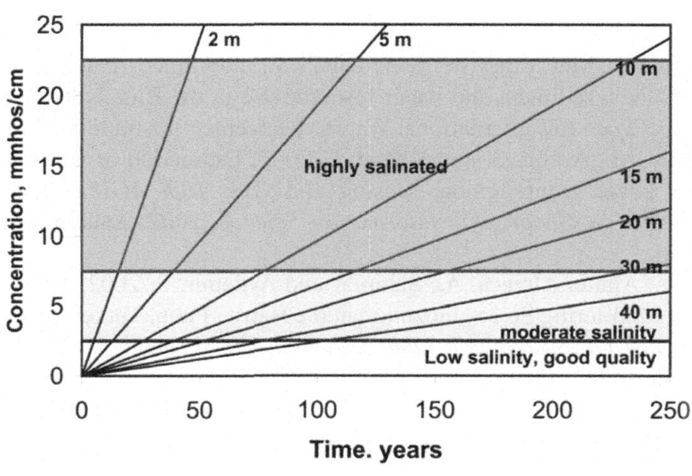

Figure 6.6 Predicted salinity concentrations in the groundwater for the Harran Plain for wells extracting water at depths ranging from 2 m to 40 m. Well depths are given as well as the water quality

China Plain. Thus, dissolved salts in the irrigation water will still be added to the aquifer. The advantage of using groundwater for some of the irrigation and keeping the groundwater table low is that salts can be leached out of the root zone, but eventually the salinity will increase in the groundwater. This is shown in Fig. 6.6 where the increase in salinization is directly related to the depth of the well that the water is being extracted from. Deeper wells allow for a greater volume of groundwater to dilute the salt and thus, a slower increase in salt content. For a well that pumps from a depth of 40 m, it will take approximately 100 years for the irrigation water from the groundwater to change from good to intermediate quality. On the other hand it will take less than 12 years for a 2 m deep well to become highly salinized. It can be easily demonstrated that by decreasing the surface water irrigation efficiency (i.e. more recharge) the aquifer salt level will increase at a greater rate.

Water price increases, or metering are measures that are often suggested for water conservation (Moench, 2001). In the case of Luancheng County and similar situations this might not be appropriate, since, in the face of increased prices, farmers will take steps to reduce the amount pumped and, as we have shown, this will not lead to any real water saving. Rather, what is required is a change in land use. Whether this will result in higher prices is debatable. Increased prices targeted at industrial water-users will probably be more effective. On the other hand in situations where salinization is a problem caused by rising groundwater, as in the Harran Plain, reducing water use will reduce the rate at which the groundwater becomes salinized.

References

Awulachew, S.B., McCartney, M., Steenhuis, T.S., and Ahmed, A.A. 2008. A review of hydrology, sediment and water resource use in the Blue Nile basin, *IWMI Working Paper 131* International Water Management Institute Sri Lanka.

El Bastaway, M., Arafat, S. and Khalaf, F. 2007. Estimation of water loss from Toshka Lakes using remote sensing and GIS. *10th AGILE International Conference on Geographic Information Science.* 2007. Aalborg University, Denmark.

Çullu, M.A., Ahmet Almaca, A., Şahin,Y. and Aydemir, S. 2002. Application of GIS for Monitoring Soil Salinization in the Harran Plain, Turkey. *International Conference On Sustainable Land Use And Management.* 2002. Çanakkale. Available at: http://www.toprak.org.tr/isd/can_50.htm.

Kendirli, B., Cakmak1, B. and Ucar, y. 2005. Salinity in the southeastern Anatolia project (GAP), Turkey: Issues and options. *Irrigation and Drainage*, 54, 115–122.

Kendy, E., Gerard-Marchant, P., Walter, M.T., Zhang, Y., Liu, C. and Steenhuis, T.S. 2003. A soil-water-balance approach to quantifying groundwater recharge from irrigated cropland in the North China Plain. *Hydrological Processes,* 17, 2011–2031.

Kendy, E., Zhang, Y., Liu, C., Wang, J. and Steenhuis, T. 2004. Groundwater recharge from irrigated cropland in the North China Plain: case study of Luancheng County, Hebei Province, 1949–2000. *Hydrological Processes,* 18, 2289–2302.

Levine, G. 1982. Perspectives on integrating findings from research on irrigation systems in Southeast Asia. *Teaching and Research Forum Workshop Report 26.* Agricultural Development Council (Bangkok, Thailand and New York).

Mehanni, A.H. 1998. The influence of depth on salinity of water table on the salt levels in the duplex red-brown earths of the Goulburn Valley of Victoria. *Australian Journal of Experimental Agriculture* 28, 593–597.

Moench, M. 2001. Groundwater: potential and constraints *2020 Vision for Food, Agriculture and the Environment. Focus No. 09.* Brief 08. Available at: http://www.ifpri.org/2020/focus/focus09/focus09_08.asp.

Richards, L.A. 1954. Diagnosis and improvement of saline and alkali soils. *Agricultural Handbook 60.* United States Department of Agriculture, Washington. Available at http://www.ars.usda.gov/Services/docs.htm?docid=10158.

Steenhuis, Tammo S., T.S., Bilgili, A. V., Kendy, E., Zaimoglu, Z., Cakmak, M.E., Cullu, M. A., Ergezer, S., 2006. Options for sustainable groundwater management in arid and semi arid areas, in *Proceedings of the International Conference on Sustainable Development and New Technologies for Agricultural Production in GAP Region*, May 29–31, Harran University Sanliurfa Turkey.

Chapter 7
Alternative Regional Water Management for Conflict Resolution in the Middle East: A Case Study of Jordan

Zeyad Makramreh

Introduction

The Mediterranean basin is one of the most water-scarce regions in the world. Water shortages are likely to become the most limiting resource for food production, economic and social development in this area. Water shortage, accompanied by high population growth and successive periods of drought, is becoming an increasing cause of stress and is likely to have a significant impact on the future political and economical framework of the Middle East.

The scarcity of water and the high cost of its development have long been recognized in arid regions where neither surface water nor renewable fresh groundwater is sufficient. The demand for water to serve expanding population growth continues to increase, while almost all fresh and renewable sources of water such as rivers, streams, lakes, and groundwater, which are known as conventional water, are shared between various countries of the Middle East. The Jordan River Basin, which is shared between Jordan, Palestine and Israel, depends mainly on surface and groundwater resources in satisfying these countries' requirements. Currently, most of the conventional water resources in these countries have reached their full rate of utilization and have been fully developed by conventional measures such as constructing dams and drilling wells.

Jordan is considered among the 10 most water-scarce countries in the world. Its water shortage problem has been exacerbated by the political situation in the Middle East and the country's high population growth. Consequently, Jordan is facing a future of very limited water resources (among the lowest in the world on a per capita basis). Available water resources are projected to decline from around 140 cubic meters per capita per year for all uses in 2008 to only 90 cubic meters per capita per year by 2025, putting Jordan in the category of having an absolute water shortage.

Based on projections of available water quantities, the gap between supply and demand from all sources is increasing annually. This situation cannot be maintained without endangering sustainable development. It shows the necessity for adopting a long-term water plan and alternative scenarios for water development that consider both demand management and non-conventional water resources, in order to decrease the gap between supply and demand.

There are a number of solutions that may be considered for water resources in Jordan and other Middle East countries. The development of non-conventional water resources will become a key measure in this century for sustaining economic development. Priority in water resource development in Middle Eastern states is still given to developing each country's own resources, not only conventional fresh-water resources but also non-conventional water resources. The potential contribution of seawater and brackish water to meeting the anticipated water demand, particularly in Jordan, Palestine and Israel, will be a unique initiative. Long-term development of non-conventional water resources in the region, particularly Jordan, must overcome the financial constraints and political complications in the Middle East. In this context, there is a recent history of optimism within the framework of the peace process. Although limited, there are signs of cooperation over multinational water development in the Jordan River Basin.

The objective of this chapter is to provide an overview of the water situation in the Jordan River Basin with an emphasis on Jordan. Water issues are presented as a model for multinational cooperation in conflict minimization, and an alternative regional management strategy for the water scarce countries of the eastern Mediterranean.

Regional Dimensions of Water Shortage

The Mediterranean basin is already one of the most water-scarce regions in the world, and conditions in the region are only expected to get worse. Water scarcity has been a source of stress since history began in this region, and the politics of water is probably of greater concern than anywhere else in the world. The water shortage is provoking situations of conflict that must be resolved and avoided. These conflicts can be of a national or an international nature, and may even have been caused by the use of this vital resource as a weapon of war and domination (Izquierdo 2003). Any regional resolution must deal with water crises if it is to be economically efficient, ecologically sustainable, and politically acceptable. This section discusses the political implications of water shortage in the Middle East and the role of peace in promoting regional water management and development.

Water Shortage and Conflict

Historically, water resources have been the origin of most political conflicts in the Middle East and North African countries. Figure 7.1 shows the geographical extent of the eastern Mediterranean region. The main watershed basins in this region are the Nile, the Euphrates and the Jordan. All three basins have experienced periods of water-related conflict and cooperation (Shuval 1992, Klaas 2003). The Jordan

River is tiny compared to the others, but the danger of conflict over its water is just as great as and even more obvious than the other two. The Jordan is the most intensely developed – with almost every drop of water planned for – and the most intricate politically, including as it does five distinct states and territories that are trying to reach agreements to end their decades-old cycle of violence (Libiszewski 1995, Darwish 1994). Water-related conflict helped form the borders of the modern states of Israel, Jordan, Lebanon, and Syria and has exacerbated tensions between the Israelis and the Palestinians.

Water was an early weapon deployed in the Arab-Israeli conflict. The water policy of Israel concentrated on restricting the consumption of domestic and irrigated water so as to provide the Israeli population with a higher proportion of water for consumption and to punish the Palestinian population (Yakhin 2006). Israel has had de-facto control of the Yarmouk River since the Six Day War that stopped Syria and Jordan from diverting the headwaters. Also, in the south of Lebanon, despite the retreat of its army, Israel has attempted to maintain control of water consumption from the Hasbani and Wazzani, tributaries of the Jordan River. In the Golan Heights region, the military occupation is also related to the water situation, since many Israeli politicians have refused to return the Golan Heights to Syria, using the excuse that they need water from that area (Izquierdo 2003).

Water policies in the Jordan valley Basin have, for generations, been at the forefront of the Middle East's longest conflict (Kliot 1994). Few agreements have been reached about how the water should be shared, and most are seen as unjust. Upstream countries believe that they should control the flow of the rivers, taking what they wish if they can get away with it, as in the case of Turkey. On the other hand, downstream countries, which are often more advanced and militarily stronger, have always challenged this assumption: Egypt and Israel, for example, considered this a recipe for confrontation. In this context, international law is not clear on the right of upstream countries to control either surface or groundwater.

Regional competition within individual states can also cause conflict, since wherever there is shortage, distribution is always controversial, and in the Mediterranean Basin water is becoming a precious commodity. Competition among different sectors of consumption is also increasing. Currently, close to 70 percent of fresh water is still used for agriculture, even in the driest areas in the region. The water conflict between states, regions and social sectors only has one valid solution: to find a confluence of interests so that water management is appropriate for everyone. A path must be forged toward a new water culture: a global and cooperative management of the basins on the part of all the sectors involved and the environment affected. Without a cooperative management model, water will continue to be used as a military and political tool, and conflicts over its distribution will not cease.

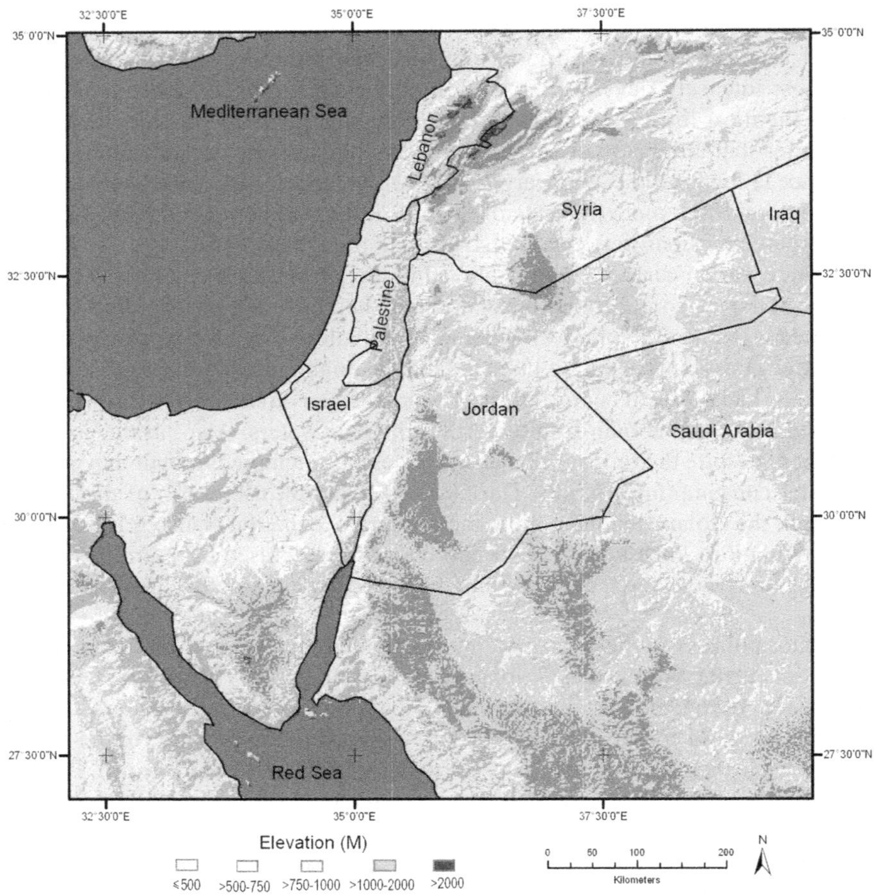

Figure 7.1 The geographical extent of the eastern Mediterranean region

The Peace Process and Regional Cooperation

With the exception of a few underground aquifers, water resources cross borders and in some cases include many sovereign states in their watershed basins. Surface water commonly crosses or forms an international border; aquifers commonly underlie a border. Besides the inter-basin surface water, trans-boundary groundwater also exists in the region. Examples include the Disi Aquifer, which underlies the border of Jordan and Saudi Arabia, and the Mountain Yarkon-Taninim Aquifer, which underlies Israel and Palestine (Shuval 2000). Political conflicts over international water resources tend to be particularly contentious because of water's preeminent role in survival. The intensity of a water conflict can be exacerbated by a region's geographical, geopolitical or hydro-political landscape (Moore 1993). Water conflicts are especially bitter where the climate is arid, and the riparians of regional waterways are already engaged in political confrontation, or where the population's water demand is approaching or surpassing annual supply (Potter et al. 2007).

International water is a subject of intense debate, with the discussion dominated by international lawyers and diplomats rather than by social or physical scientists. In the Middle East, the basic principle for sharing water remains that of equitable use. This implies that the ways specific bodies of water are shared must be negotiated to fit the physical, economic, and social context of the parties involved (Brook 1996). The right of parties to specific quantities and quality of water remains a contentious issue. In these circumstances, it might be helpful to shift attention from rights aimed at the supply side to rights guaranteeing certain levels of demand.

International law is not clear on the issue of shared water courses, rivers or transboundary aquifers. Although international law applies most directly to surface water, each of the principles used in dealing with surface water applies also to underground water. A model treaty for internationally shared aquifers has been drafted (Hayton and Utton 1989), but it has not yet been extensively discussed by politicians. Discussions about international waters, including those in the Middle East, typically conclude with a call for basin-wide or aquifer-wide commissions to manage them as a unit; such schemes need to establish trust among the nations concerned in order to consider joint regional management of shared water resources (Elmussa 1993). The general proposition of 'water wars' ignores the wide range of options available for overcoming water scarcity that can relieve the pressure much less expensively and with much less risk than would be incurred in military conflict (Hillel 1994). There are simply better alternatives than war, although they may not be politically easy or free of conflict (Homer-Dixon et al. 1993). The slow-progress approach toward regional management in the Middle East is not intended to preclude cooperation through prior notification of changes in river regime or specific joint institutions. Nor does it exclude the possibility of true joint management in cases such as the Mountain Aquifer, where Israelis and Palestinians really have no other alternative (Kally and Fishelson 1993). Even in these cases, slow cooperation with parallel but not joint institutions on either side of the border would probably be more successful than attempts to move quickly to establish regional institutions.

The same characteristics of water resources that fuel conflict can, if managed carefully, induce cooperation, even in an environment of hostility. After the Gulf War in 1990 a new phase of the peace process was started in the Middle East and finally made possible, if with limited progress, regional cooperation between Arabs and Israelis. During the bilateral negotiations between warring countries it became obvious that water was at the heart of the negotiations since other issues that had obscured water for years proved to be of lesser significance to the parties of the dispute (Haddadin 2001). During the peace talks, agreements were reached on a wide range of principles and projects. These agreements had major components dedicated to jointly and collectively solving the regional water shortage.

There have been several changes in the Israel-Arab situation since the beginning of the peace process. As a result of integrated hydrological studies of the Jordan River system, it is now possible to conceive of comprehensive development strategies that will not only be technically and economically feasible but also politically desirable (Beaumont 2002). Currently, one benefit of the peace process in water management issues is a feasibility study for a regional cooperative project between Jordan, Palestine and Israel with financial aid from the World Bank, namely the Red-Dead-Sea Conduit (RDSC). Such projects have been conducted without threatening new political conflicts but rather promoting political and economic development in the region.

Regional Characteristics

The Eastern Mediterranean region, especially Jordan, Palestine and Israel, have a Mediterranean-type climate, which is characterized by hot, dry summers and cool winters with short transitional seasons predominantly in the northern, central, and western parts of the Jordan Rift Valley. The eastern and southern parts of the region have a semi-arid to arid climate. Winter begins around mid-November and summer begins around the end of May. Rainfall occurs mainly during the winter months. Natural replenishment of water resources varies greatly and exhibits major changes within a relatively small distance across the region.

These countries experience extreme seasonal variations in climate. Large rainfall variations also occur from year to year. Average rainfall decreases from west to east and from north to south, ranging from 1,200 mm/yr at the northern tip of the mountains to less than 50 mm/yr in the desert areas. Rainfall of less than 200 mm/yr constrains development of agricultural land use and rangeland production in about half of the area on the western side, and 90 percent of the area on the eastern side of the Jordan Rift Valley. 'Wet' and 'dry' periods usually last several years and make managing the region's precious water resources an even greater challenge. As shown in the following sections, where the water supply-demand in watershed of the Jordan is discussed in relation to population growth, these climatic factors affect water-use practices, policies, and expectations.

Population Growth and Water Demand

Most countries in the Middle East region, especially the Arab countries, are experiencing rapid population growth with average growth rates of 2.5 percent per year. Other sources of stress on water resources come from rapid urbanization which increases the demand for high water quality without diminishing the demand for irrigation water, and from booming economic growth (CEDARE 2007). The list of water scarce countries in 1955 was seven, including three Middle Eastern countries: Bahrain, Jordan and Kuwait. In 1990, 13 more were added, among them eight from the Middle East: Algeria, Palestine, Qatar, Saudi Arabia, Somalia, Tunisia, United Arab Emirates, and Yemen. It is anticipated that by 2025 another seven countries from the Middle East: Egypt, Ethiopia, Iran, Libya, Morocco, Oman and Syria will be added to the list. This means that by the year 2025 some eighteen countries in this troubled region will suffer from water shortages (Berman and Wihbey 1999).

Jordan's population increased ten-fold between 1952 and 2006, reaching approximately 6 million and is projected to reach about 10 million in 2020. Jordan's annual per capita water availability in 1990 was 327 cubic meter (m^3) and in 2006 about 160 m^3, some 673 m^3 and 840 m^3 below the bottom limit of water crisis respectively (the bottom line for water shortage is 1,000 m^3 per year). In addition, Israel's population is projected to grow from 4.7 million in 1990 to about 8 million in 2025. By that time, because of their higher birth rate, the population of Palestinians in the West Bank is likely to reach about 7 million. This means that the same water resources will have to be shared by at least double the present population.

Sustainable water supply in Jordan is limited, whereas demand is rising rapidly. Renewable freshwater resources are of the order of 750–850 MCM with approximately 65 percent derived from surface water and 35 percent from groundwater sources (Nortcliff et al. 2008). Current demands for water are of the order of 955 MCM, of which 450 MCM is derived from surface water while the remainder comes from renewable and non-renewable groundwater. Figure 7.2 shows the projected trend of water deficit in Jordan up to 2020 based on water demand and supply. The present annual water demand amounts to 10 percent of the annual total rainfall of the country. The estimated deficit between the supply and demand in 2020 is projected to be about 400 MCM.

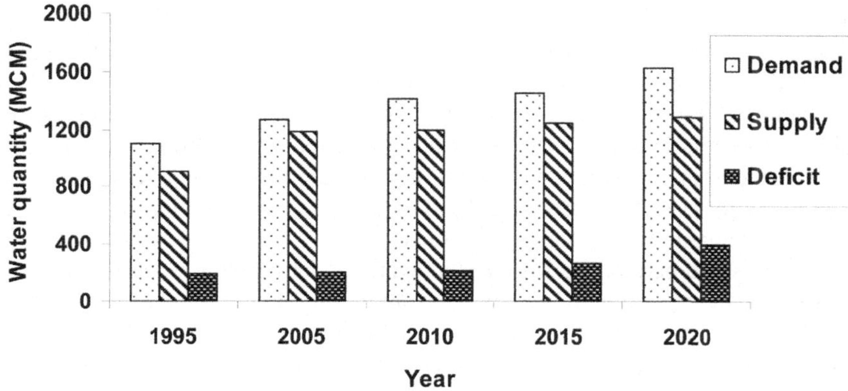

Figure 7.2 Water consumption of different sectors in Jordan

Almost all the economically viable surface water resources in Jordan have been harnessed, mainly for irrigation purposes. Agriculture is the main consumer of water with about 67 percent of the total use, followed by the domestic sector with 29 percent, and only 4 percent for the industrial sector (Figure 7.3).

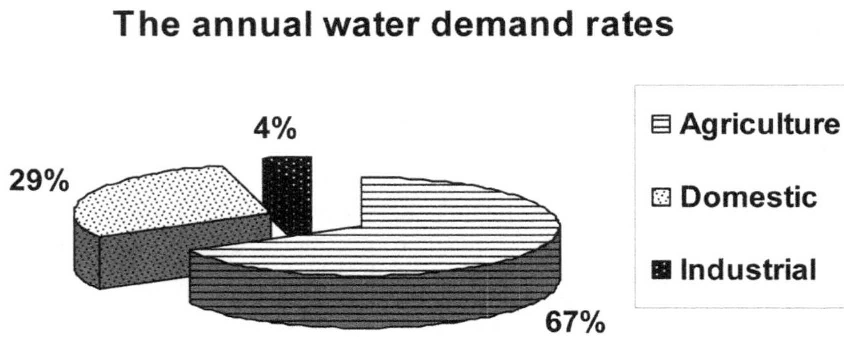

Figure 7.3 The projected trend of water deficit based on water demand and supply in Jordan (MWI)

To meet the deficit between supply and demand the groundwater aquifers have been mined at a rate about 160 percent of their sustainable yield (Hussein 2002). Still, Jordan is suffering severely as a result of water-rationing. The pressures on water are most marked in Amman, the capital, where the vast majority of households receive water only on one or two days per week (Potter et al. 2007).

Jordan is classified as one of the most limited water resource countries in the world. This is combined with one of the lowest per capita water availability ratios of any country, with about (140 m^3/yr) in 2007 (Nortcliff et al. 2008). This amount is projected to drop as a result of population growth to about (90 m^3/yr) by 2025. Therefore, according to the water stress index, Jordan is classified as having absolute water scarcity (Abdel Khaleq and Dziegielewski 2004).

Table 7.1 shows the total amounts and percentages of projected water use by different sectors in Jordan. The relative share of municipal and industrial supply is expected to increase from 33 percent of total water use in 2005 to 51 percent in 2020. The increase in domestic and industrial water consumption will affect the quantity available for the agricultural sector.

In the same period, the Ministry of Water and Irrigation model forecasts an increase in the recycling of wastewater from 67 MCM/yr in 1998 to 232 MCM/yr in 2020, with a parallel reduction in groundwater abstraction by 122 MCM/yr. This recycling is necessary to reduce the current overexploitation and to protect aquifers from salinization. However, even if all the planned projects and measures included in the Ministry of Water and Irrigation's investment program were to

Table 7.1 The total amounts and percentage of projected water use by sectors in Jordan (MWI, 2002)

Year	Domestic and Industrial (MCM)	Percent	Irrigation (MCM)	Percent	Total (MCM)
1985	138	21.6	501	78.4	639
1990	212	24.4	657	75.5	870
1995	272	31.0	606	69.0	878
2000	276	33.8	541	66.2	817
2005	357	32.3	750	67.7	1107
2010	473	38.8	746	61.2	1219
2015	575	45.0	704	55.0	1279
2020	647	49.3	665	50.7	1312

Source: MWI.

be implemented, which is unlikely, the net increase in available supplies by 385 MCM/yr would not keep pace with the increase in water requirements of 442 MCM/yr.

Water Resources

The water resources of the Jordan River basin, particularly in the countries of Jordan, Palestine and Israel, depend mainly on precipitation and surface water. The Mediterranean watershed includes the coastal plain, parts of the mountain belt, and the desert area. Water resources in Jordan consist primarily of surface and groundwater, with treated wastewater being used on an increasing scale for irrigation, mostly in the Jordan Valley.

The average total quantity of rainfall of Jordan is approximately 7,200 MCM/yr, varying between 6,000 and 11,500 MCM/yr. Renewal of water resources depends on the overall amount of precipitation and is affected by temperature, evaporation, rates of runoff and groundwater recharge. On the western side of the Jordan Rift Valley, approximately 70 percent of the total precipitation is lost through evaporation and 30 percent is usable: 5 percent is runoff, and the remaining 25 percent is recharge groundwater.

On the eastern side of the Jordan Rift Valley, 90 percent of the total precipitation is lost to evaporation, 5 percent is runoff, leaving only 5 percent for groundwater recharge. Of the 5 percent that recharges the groundwater, a portion is eventually discharged as base flow into streams or springs which are then classified as surface-water resources. The water is stored in the groundwater reservoirs and is potentially available for withdrawal from wells.

Surface Water Resources

The Mediterranean watershed on the western side of the Jordan river includes the coastal plain and parts of the mountain belt and the Negev. The streams generally have small watersheds with headwaters in the western mountains. Many of the streams are affected by water supply diversions and wastewater discharges. On the eastern side of the Jordan River the main watershed catchments are the highland mountains, the river basin and the desert watershed. The location and boundaries of the major surface watersheds in Jordan are shown in Figure 7.4.

Watershed size is a poor indicator of relative flow because of the extreme differences in climate across the region. Few streams outside the Jordan River watershed have adequate base flow from groundwater and springs to flow throughout the year. Many streams of the Mediterranean and Dead Sea watersheds flow throughout the rainy season and are dry during the summer. Streams of the Wadi Araba and Desert watersheds typically flow only in response to winter storms. Peak flows typically occur during February and March, lagging behind the peak precipitation period by about one month. This lag time is due principally to the balancing of extreme moisture deficits in parched soils and plants after the dry season. Floods may also occur following intense storms in the spring and fall.

The three major rivers in Jordan are the Jordan, Yarmouk and Zarqa, all of which have become highly unreliable for water supply. For the Jordan and Yarmouk rivers, this is due to upstream diversion and over-utilization by Israel and Syria, leaving Jordan with whatever remains. The Zarqa river basin has been severely affected by industrial pollution, overexploitation and drought.

The Jordan River watershed has the largest water yield in the region and provides most of the usable surface-water supply. It includes the Mountain Belt, the Jordan Rift Valley and Escarpments, and the Jordan Highlands and Plateau. The Jordan is a multinational river, flowing through Lebanon, Syria, Israel, Palestine and Jordan. Since construction of the Al-Wuheda dam was completed in 2005/2006, it has become overdeveloped except for a winter flow in its largest tributary, the Yarmouk River, which forms the present boundary between Syria and Jordan for 40 km before becoming the border between Israel and Jordan (Huang and Banerjee 1984). Currently Israel uses as much as 90 percent of the stream water from the upper Jordan River, and consequently Jordan's water problems have been exacerbated by Israel's policy of denying Jordan the right to utilize and develop the water of the Jordan River within its borders.

The streams generally have small watersheds with headwaters in the western mountains. Annual stream flow generally declines from west to east as the distance from Mediterranean moisture sources increases, and from north to south with increasing temperature and evaporation. Stream flow is typically higher on the western side of the Mountain Belt, due to temperature and orographically-induced precipitation, and decreases on the eastern side of the Mountain Belt descending into the Jordan Rift Valley (Naff and Matson 1984). River flows are generally of a flash-flood nature, with large seasonal and annual variation. Annual base

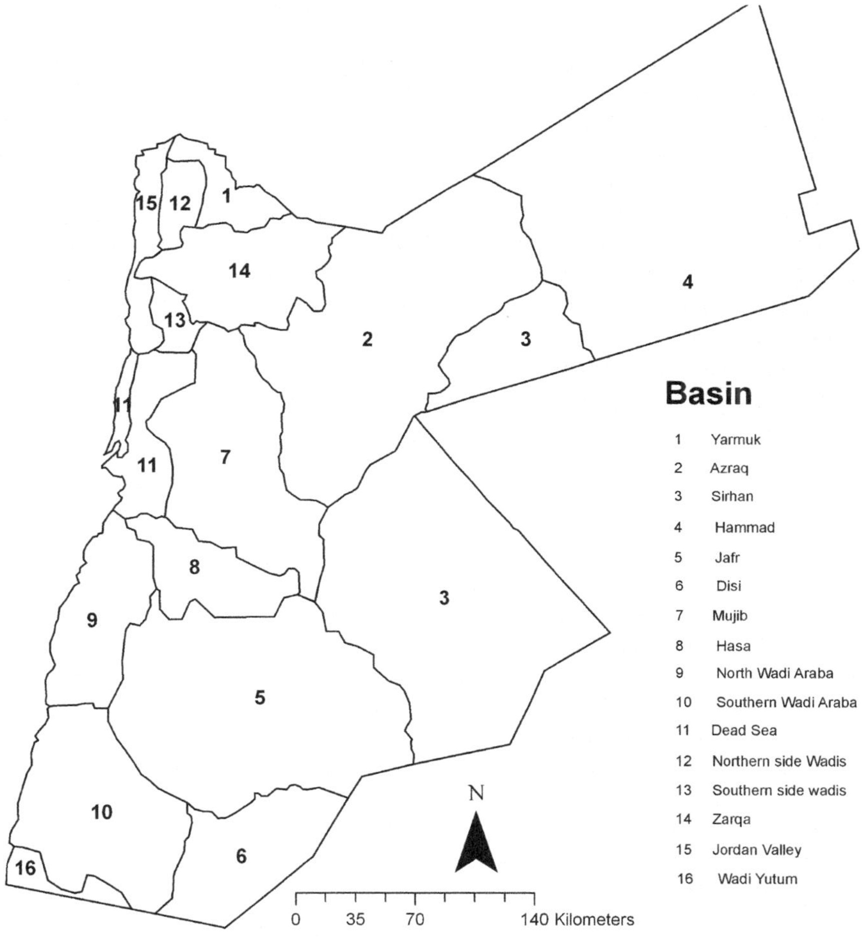

Figure 7.4 Surface water basins in Jordan

flows, whose volume is estimated at 540 million cubic meters, vary by at least 15–20 percent, depending on rainfall patterns, with a return period of five years (Huang and Banerjee 1984).

Groundwater Resources

Groundwater is derived from two sources: fossil water, which receives only a very small amount of recharge; and recent and renewable water. Fossil aquifers are non-renewable and are found mostly in the southern and eastern parts of the region. Recent and renewable recharge is derived naturally from precipitation, or from streams, wadis, lakes, ponds, or other catchments that seep through soil into the aquifers. The most productive aquifers of the region are in quaternary sand and gravel of the Coastal Plain, cretaceous limestone in the Mountain Belt, the eastern and western escarpments of the Jordan Rift Valley and Jordan Highlands, basalt of the Jordan Highlands and Plateau, and sandstone of the South Jordan Desert. Figure 7.5 shows the major groundwater basins in Jordan.

The major potential groundwater aquifers are found in the pervious sequences of the basalt system of the Pleistocene, the Rijam formation of the Lower Tertiary, the Amman-Wadi Sir formation of the Upper to Middle Cretaceous, the lower Ajlun formation of the Middle Cretaceous, the Kurnub Zarqa formations of the Lower Cretaceous, and the Disi formations of the Palaeozoic age. The shallow aquifer systems of basalt-Rijam form a locally important aquifer in the central part of the Jafr and Al-Azraq-Wadi. It has been intensively exploited for the municipal

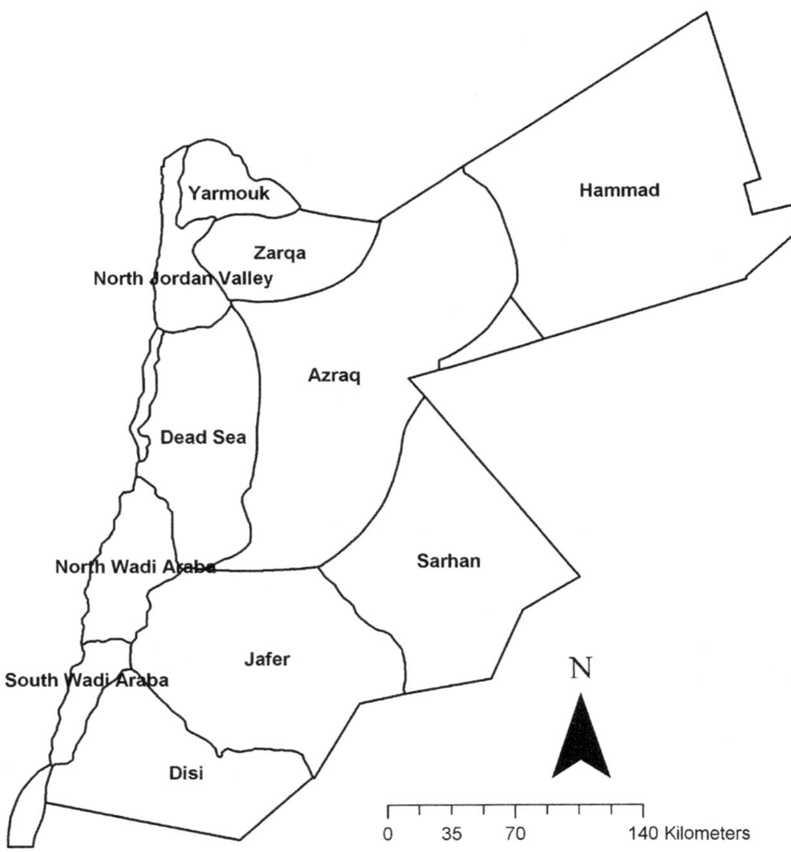

Figure 7.5 The major groundwater basins in Jordan

water supply and its annual abstraction has exceeded the safe yield, lowering the piezometric head and increasing the salinity of the water.

The most important aquifer system in Jordan is the Amman-Wadi Sir, which consists of limestone, silicified limestone, chert, sand limestone, and sandstone of the Upper to Middle Cretaceous age. It has been excessively exploited in the northern part, while in the southern part it is economically the most important of the country's aquifers, because the good quality water it contains is less than 500 mg of TDS per liter.

On the other side of the country the deep sandstone aquifer structures are the Kurnub/Zarqa of the Lower Cretaceous age and the Disi of Palaeozoic age. They are unconformable, being separated by a less permeable layer of sandstone, siltstone, and shale. The Kurnub formation intercalates frequent argillaceous layers in the south, while the Disi is composed of massive and rather homogeneous arenaceous layers. Groundwater in these aquifers is mostly non-renewable because of limited groundwater recharge through the small outcrop areas. The quality of the groundwater in the Kurnub-Zarqa system varies from fresh to brackish. However, excellent quality water with low salinity is found in the Disi aquifer in the southern part of the country. The development potential of the Disi aquifer has been estimated as about 100–200 MCM/yr for a period of over 50–100 years (Mohsen and Al-Jayyousi 1999). Currently the Disi aquifer is used to supply the city of Aqaba with water for domestic purposes and has been used for agricultural purposes for three decades (Jaber and Mohsen 2001).

Groundwater is the most important source of water supply in Jordan and is presently used for municipal, industrial, and agricultural purposes, providing more than half of the total water consumption. More than 90 percent of Jordan's population is now supplied with drinking water from springs and groundwater wells.

Data from the monitoring program of the Water Authority of Jordan (WAJ 2005, open files) show that since the early 1980s, when over-exploitation began, major aquifers in Jordan have been experiencing declines in their water levels. The safe and abstracted yield from the groundwater basins in Jordan is shown in Table 7.2.

Current exploitation of these groundwater resources is at maximum capacity and in some cases exploitation is well above what is recognized as a safe yield. The safe yield of renewable groundwater basins is around 275 MCM/yr (NWMP 1977). In the year 2000, about 411 MCM/yr (Al-Halasah 2003) was abstracted from renewable groundwater, but the actual extraction in 2007 was around 520 MCM/yr (Salameh 2008). From this value 80 percent is renewable and 20 percent is from non-renewable groundwater resources. This level of groundwater consumption corresponds to more than 160 percent of the total aquifers' sustainable yield. The water levels of the major aquifers are dropping at rates ranging from 30 to 120 cm/yr (WAJ 2005, open files). At the present extraction rates, the major aquifers are expected to be irreversibly depleted within 10–30 years (Salameh 2008).

Table 7.2 The safe and abstracted yield from the main groundwater basins in Jordan measured in millions per cubic meter

Groundwater Basin	Safe Yield	Water abstracted	Balance	Percentage abstracted
Yarmouk	40	43.3	-3.3	108
Jordan Valley	21	27.9	-6.9	133
Azraq	24	59.3	-35.3	247
Amman-Zarqa	87.5	138.7	-51.2	158
Serhan	5	3.8	1.2	76
Hamad	8	0.9	7.1	11
Dead Sea	57	89.3	-32.3	157
Disi	125	82.1	42.9	66
North Wadi Araba	3.5	6.7	-3.2	193
South Wadi Araba	5.5	17.4	-11.9	316
Jafr	9	24.8	-15.8	276

Alternative Perspectives for Water Management

Most of the conventional water resources in Jordan have been fully exploited or developed by conventional measures such as constructing dams and drilling wells. A number of solutions to water resource scarcity can be considered, ranging from agricultural to technological, economic and public policy solutions, but they all fall under the two basic categories of water supply-demand management, and development of non-conventional water resources. The development of marginal non-conventional water resources will therefore become a key measure in this century for sustaining economic development in the countries of the Middle East (Murakami 1995, InWEnt 2005). The next section reviews the most feasible water management and non-conventional water development alternatives that can be used in Jordan.

Water Management Options

The supply-demand management option involves increasing the supply of water, particularly from deep aquifers or inter-basin surface water transfers. On the other hand, the option to decrease demand using market and public policy forces may lead to more efficient allocation of water. Another option is the use of various water harvesting techniques and the re-allocation of water from one sector to another. The next section reviews the most feasible water management options for Jordan.

Development of Current Water Resources

Since the beginning of the 1960s, several dam schemes have been implemented in the main water catchment areas in the mountains along the Jordan Valley Basin in order to utilize the surface water resources mainly for irrigation purposes. Currently Jordan collects about 50 percent of the total surface runoff water resulting from rainfall which is about 677 MCM/yr, while the storage potential of all dams in Jordan is around 327 MCM. The most recently built reservoir is the Al-Wehdah dam on the Yarmouk River, which was completed in 2005/2006. Theoretically, this dam has added about 110 MCM of storage capacity (THKJ 2004) but the reservoir did not fill up in the 2007–2008 season due to over-utilization of the Yarmuk River and the influence of drought.

Unfortunately, no unused surface water remains to be developed in the Jordan Basin. Also, because of climatic, topographic and hydro-geotechnical conditions, the construction of new storage dams in Jordan is extremely costly and not economically feasible (Potter et al. 2007). Investment in storage dams in Jordan can only be justified for the supply of municipal and industrial water or for the irrigation of high-value, high-yield crops using water-conserving technologies.

Until now, the government's strategy has been to use groundwater resources for municipal, industrial and agricultural purposes and to utilize surface water primarily for irrigation. Domestic water supply depends exclusively on the groundwater supply, owing to its better quality and the lower cost of investment compared to using surface water resources. Accordingly, because of its high population growth, most of Jordan's groundwater has been exploited. The Disi aquifer, about 350 km south of Amman is the only remaining significant aquifer, and the most valuable in Jordan. It is, however, a typical fossil aquifer, with an estimated safe yield of about 100 million cubic meters per year over a 100 year period. It also represents Jordan's last substantial unexploited fresh water resource. Currently, the Ministry of Water is carrying out the largest groundwater development project in Jordan which should be completed by 2013 in order to transfer the water from the Disi aquifer in the south to the main urban centers in the center and north of Jordan. When this aquifer is fully developed there will be no other alternatives for water resource development except the use of non-conventional options.

Water Harvesting

The basic principle of water harvesting techniques is to collect rainfall over a relatively large area and use it to irrigate a small portion of that area. However, for agricultural use this technique is only applicable in selected places, depending on soil structure and composition. Moreover, the collected water is distributed over large distances and therefore only available for local use. This technique, if extensively applied, will add 30 to 50 MCM/yr or around 5 percent to the water supply of Jordan (Salameh and Bannayan 1993).

Another form of water harvesting consists of collecting rainfall from the roofs of houses and storing it in cisterns for domestic use. This option produces only small quantities of water but it is potable. Even in areas of low rainfall such as Jordan, it is possible to design low-cost systems with cisterns scaled to household use. Rainfall over a typical 80 square meter roof in Amman's densely populated suburbs would account for 32 cubic meters yearly and would supply 17 percent of the water consumption of a six-person household (Schiffler 1993). The greatest obstacle is not designing the systems but convincing people unused to this technique that the stored water is, indeed, potable. Jordan's Environmental Action Plan has suggested reviving this technique by requiring all home-owners to install water collecting facilities on their roofs.

Water Conservation and Demand Management

Conservation of water, including both increases in the efficiency of existing uses and changing use-patterns, has always been a major element of Middle East water strategies. Private firms and public bodies must begin to look at reductions in the use of water as a source of supply, one that is equivalent to and in many ways better than the exploitation of new primary sources. Although none of the countries under consideration comes close to maximizing the economic, much less technical potential of efficiency in water use, the dominance of irrigation requires special attention. For example, Jordan and Israel are generally regarded as models of efficiency in irrigation, particularly because of the development and adaptation of drip irrigation systems (Molle 2006).

Demand management can take many forms, from measures to diminish loss, to technical measures for improving the efficiency of water use at the system level or at the user end by controlling or reallocating water consumption among various sectors of utilization. In terms of policy, measures can be direct, aimed at prescribing and/or rationing water allotments by administrative order; or indirect, influencing voluntary behavior through market mechanisms, financial incentives, or public awareness programs.

Demand-side management also encompasses the institutional arrangements supervising the water sector which often have a considerable influence on allocations and consumption patterns (Berkoff 1994). The greatest potential for reducing water demand is probably the reallocation of water among the various uses and sectors of consumption. Reallocation means shifting water from those uses and sectors which show a low added value per unit of water consumed to those of primary social need or with higher water productivity. In other words, this approach calls for restructuring the economy away from heavily irrigated agriculture towards other sectors, in particular domestic consumption and industrial uses.

Taking into account the real importance of agriculture for the overall economy, re-sizing this sector in at least some of the countries concerned appears less unthinkable than at first assumed. For example, Jordan is no longer an agrarian society, as one might assume. In fact, the country has a high level of urbanization (more than 80 percent), and the average contribution of agriculture to the GNP is less than 6 percent. Nevertheless, some long-term adjustments aimed at gradually reducing irrigated agriculture and promoting water-saving industrial and commercial activities seem inevitable (Schiffler 1994). According to some researchers, the socio-economic impact of water reallocation from the agricultural sector for rural communities is a much more acceptable social and economic policy than is recognized (Haddadin and Shteiwi 2004). Other analysts believe that water-scarce areas, such as the Jordan Valley countries, will have no alternative but to turn to external sources as soon as possible. These analysts argue that even a higher level of end-use efficiency and a total removal of freshwater from agriculture would be insufficient to solve the problem. Therefore, development of non-conventional water resources is essential to overcoming the social, economic and political impacts of water shortage.

Development of Non-Conventional Water Resources

Conventional alternatives have the highest priority in water-resource planning where there are still renewable fresh waters to be developed without creating any inter-state conflicts. However, after exploiting all the renewable fresh water resources within their national boundaries, Jordan, Palestine, and Israel have no choice except to develop trans-boundary and/or non-conventional water resources (Salameh 2008, Jaber and Mohsen 2001).

As we have seen, the potentially exploitable renewable water resources are reaching their limits in Jordan as a result of increasing demand especially in urban centers. Taking into consideration climatic conditions and the limits of conventional water resources, the development of non-conventional water alternatives is becoming imperative to supply fresh potable water to the growing population. Fortunately, the hydrology and hydrogeology of Jordan provide several possibilities for potential water development, such as wastewater treatment and desalination of both brackish water and seawater.

Wastewater Treatment

A promising source of non-conventional, technology-intensive supply is the purifying and reuse of industrial and domestic wastewater. It is assumed that about 65 percent of the water consumed by industry and private households can be recycled if collected by sewage systems and properly treated (Assaf et al. 1993). Wastewater treatment became an important water source in the region in the 1980s, and is currently an accepted practice in countries such as Jordan, Egypt and Israel.

These countries have begun to shift from fresh to recycled water for irrigating crops in recent decades and the practice will accelerate in this century

(Nazzal et al. 2000). At present, between 25 percent and 33 percent of total water consumption in the countries of the Jordan Basin region is attributable to the private and industrial sectors. This share is likely to increase rapidly as a consequence of population growth, improvement of living standards and economic development.

The Ministry of Water and Irrigation in Jordan has an ambitious ongoing sewage treatment program which will not only have positive environmental and health impacts but will also provide for the collection and treatment of sewage in a way that effectively lends itself to the reuse of treated effluents. Significant work on sewage treatment has taken place during the last decade in Jordan, and more than 60 percent of the urban population is now being served (Malkawi 2003). Currently, treated wastewater originates from more than 19 treatment plants, with around 70 to 80 MCM/yr of treated wastewater discharged into various water bodies or used directly for restricted irrigation mainly in the Jordan Valley (Nsheiwat 2007).

The wastewater strategy plan is intended to complement the existing wastewater treatment plants in Amman, Zarqa and other cities; many other plans are underway to construct treatment plants to serve an additional 34 cities and villages in Jordan. These plants will have a combined capacity of 110 MCM by the year 2010. It is expected that the amount of wastewater used for irrigation by 2020 will reach 232 MCM as shown in Table 7.3 (WMI 2002).

Table 7.3 Past and projected treated wastewater reuse in Jordan by 2020 (MCM/yr)

Year	Jordan Rift Valley	Highlands	Total
1998	56	11	67
2005	65	41	106
2010	110	45	155
2015	123	74	197
2020	137	95	232

Desalinization of Seawater and Brackish Groundwater

The prospect of desalinating seawater and thus gaining an almost inexhaustible source of fresh water has intrigued experts for a long time. Most of the current global desalinization capacity is already installed in the Middle East: Saudi Arabia has more than 26.8 percent, Kuwait 10.5 percent, and the United Arab Emirates 10 percent. The greatest constraint to widespread use of seawater desalination is its cost. In fact, the technology remains very expensive, making it currently impracticable for most applications. However, advanced technologies are being applied today in the desalinization of both brackish water and seawater in the effort to make the process more economically and commercially feasible.

Jordan contains both seawater and brackish water. More convenient and thus more realistic in the short term is the desalinization of brackish water. Such water is relatively widespread in Jordan, either for natural reasons or because of increasing salinization of fresh supplies (Abu Qdais and Batayneh 2002). One study (JICA 2001) concluded that there is a potential for producing 60 MCM/yr of desalinated brackish water from groundwater resources in the Hisban, Kafrain, Karameh, and AbuZieghan areas. Desalinization by-products can be utilized by the existing industrial sector for salt production.

A large amount of exploitable brackish groundwater may also be stored in the deep sandstone aquifers in southern Jordan such as the Kurnub and Khreim formations, in addition to the shallow aquifers of the eastern desert, including the areas of Azraq, Sirhan, and Hamad. This brackish groundwater is situated only 100–150 km east of Amman, which suggests a potential source of water supply for the Amman municipalities if cost-effective desalinization is developed.

The quality of the brackish groundwater ranges from 1,000–2,000 mg/l to 5,000–10,000 mg/l, which is unsuitable for domestic use or irrigation. Brackish groundwater with salinity of less than 2,000–3,000 mg/l can be used directly for some crop irrigation, depending on soil conditions. However, Jordan's experience with brackish water desalinization has been fairly limited. Despite the potential cost reductions that can be accomplished through the use of relatively inexpensive solar energy ponds to power large desalinization plants, all of the plants to date have been built mainly to serve local domestic and some industrial water supply requirements.

Desalinization of seawater for municipal and industrial supply is the principal source of water in the rich oil-producing Gulf states. However, Jordan has very limited energy resources and only has access to the sea at the Gulf of Aqaba, where it has a short coastline. Several studies have been carried out in the past on the possibilities for desalinization but none of these have been shown to be economically attractive (Al-Jayyousi 2001). The saline water from the Gulf of Aqaba has a TDS of 43,000 parts per million and represents an unlimited resource of water, but in addition to the desalinization process, it would have to be transported about 350 km to Amman and even further to other cities and would involve high pumping costs.

Therefore, the desalinization of seawater or brackish water in Jordan has many economical and financial constraints. However, according to an analysis conducted by academic researchers in Jordan (Jaber and Mohsen 2001), it was concluded that desalinization is the most appropriate option for Jordan to adopt in order to alleviate water scarcity and overcome its water budget deficit.

Currently under development is a large desalinization plant at Abu Zighan which will deliver 18 MCM/yr at maximum capacity. In 2004, an important plant was built to desalinate the brackish groundwater of the Jordan Valley (Deir Allah, 15 MCM/yr which supplied Amman with 10 MCM/yr (MWI 2002)). The most important cost factor in such desalinization processes is the energy cost, which can be controlled by introducing off-peak power operations, taking into account the dominant steam-power generation with high peak demand in Jordan. Recent innovative developments in the high-molecular membrane industry could provide the necessary energy saving through the use of low-pressure reverse-osmosis modules.

However the question of funding looms large over the possible construction of desalinization plants and the cost of such plants is beyond the purchase capability of Jordan (Fisher and Huber-Lee 2005). The sponsorship of desalinization plants is an ideal opportunity for international involvement, and could pave the way for more regional cooperation in water issues between Israel, Jordan and the Palestinians. In conclusion, the development of saline water resources by desalting with reverse osmosis and other appropriate methods will play an increasingly important role in the context of the future national and regional water strategies in Jordan and other Middle East countries.

Regional Solution: Red-Dead Sea Conduit

A regional alternative scheme to overcome the water shortage in Jordan, Palestine and Israel is the Red-Dead Sea Conduit project (RDSC). The idea comes from the fact that the Dead Sea is separated from the Red Sea by a mountain ridge of modest proportions and is located below sea level, while its salinity is even higher than that of the Red Sea (Harza JRV Group 1996). The main objective of this project is to replace the water losses in the Dead Sea due to diversions along the Jordan and Yarmouk rivers. A further goal is the production of 850 MCM of desalinated water per year using the process of reverse osmosis desalination as a multi-country venture expected to contribute to regional water development.

Under this scheme, seawater would be pumped into a series of canals and reservoirs from Aqaba to the Dead Sea, and there are tentative proposals to build a pipeline to transfer this water. Water would be transferred through a pipe that would offer good conditions for hydropower generation (Beyth 2007). A total transfer of 1.5 billion cubic meters per year is planned to both supply municipal water and stabilize the level of the Dead Sea. Part of the electricity generated would be devoted to desalinating seawater and thus gaining new fresh-water supplies (Murakami 1995, Glueckstern and Fishelson 1992).The capacity of the reverse osmosis plant will be 850 MCM/yr, while the conveyance system from the Dead Sea to Amman will be 570 MCM/yr, the remaining part being supplied to Israel and the Palestinian Territories.

The planned water volume that would be transferred into the Dead Sea is close to that historically brought by the Jordan River (between 800 and 1,000 MCM/yr). The water in the Dead Sea would be maintained at a steady-state level, with some seasonal fluctuations of about 2 m, between 402 and 390 m below mean sea level, with the inflow into the Dead Sea balancing evaporation. The maximum level reached by the Dead Sea would not flood any religious or archaeological sites, nor would it trigger earthquakes, as this level is comparable with previous equilibrium levels, and would not increase reflectivity. Despite the fact that there are adverse environmental impacts associated with the RDSC project, its benefits outweigh these impacts which can be eliminated or minimized by adopting proper measures (Abu Qdais 2008).

Estimates of product water costs range from 1 to 2 US$ per cubic meter according to Wolf (1995). Another study by Murakami and Musiake (1994) cites costs of 0.68 US$ per cubic meter of water (at a price level of 1990) for a plant producing 100 MCM annually, not counting distribution expenses. These prices may only be affordable if the water is primarily used to supply tourist facilities on the shores of the Dead Sea and eventually for some very sophisticated agro-industrial complexes, but not for conventional agriculture. Total costs of the project have been estimated at 5 billion US$ (1.3 billion for the Sea Water transfer and 3.7 billion for the desalinization and freshwater transfer facilities; (Harza JRV Group 1996)). However, the RDSC project is considered the only real alternative for the future strategic development of non-conventional water resources, one that will also conserve fossil energy and local environmental resources.

Conclusions

Jordan is one of the ten most water-deprived countries in the world. Available per capita freshwater lags far behind the standard international level. Daily water consumption is also quite low and the cost of supplying water continues to rise. This extreme scarcity and the increasing cost of water supply are very serious constraints to Jordan's economic and social development. There are a variety of solutions to water resource shortages that can be considered. They all fall under two basic categories: management of current water resources and development of non-conventional water resources.

The development of current water resources will be relatively expensive and is not enough to narrow the gap between demand and supply. Water conservation will be essential to managing water resources in Jordan, especially after the renewable water resources have been exploited to their limits. In addition, rainwater harvesting options i.e. collection of water from rooftops, is an option for potable water, but it supplies only limited amounts. Once the exploitation of the fossil groundwater in the Disi aquifer has been completed, non-conventional alternatives will become the key to future water plans for developing new water resources.

Among the non-conventional alternatives are the re-use of treated wastewater and desalinization of seawater and brackish water. Recycling of waste-water can yield only marginal quantities, mainly suitable for agriculture purposes. Hence, the only remaining option for satisfying the municipal water demand is desalinization. The desalinization process will play a vital role in satisfying the ever-increasing demand for water in various sectors. The great advantage of desalinization, despite its difficulty and expense is that there are no political constraints on its development.

The history of hydro-politics along the Jordan River Basin exemplifies the best and the worst of international water relations. Shared water resources have brought Middle East countries to the brink of armed conflict but they have also been a catalyst for cooperation between otherwise hostile neighbors. Nowadays, what the Middle East faces is not so much a water crisis as a chronic problem concerning technological, political, and inter-regional management issues which is escalating to crisis dimensions. Water shortage will became a vital factor for political, economic and social development in the region. Therefore more water must be supplied for the use of these countries through various methods and the focus should be on cooperative regional and international projects within the Jordan River Basin.

In recent years, water politics and bilateral negotiations have placed a high priority on regional water management and innovative technological development combined with sound economic development. In this context, Jordan is actively involved in promoting regional cooperation through the Water Resources Working Group of the Multilateral Peace Talks. The most recent benefit of this regional cooperation is the feasibility study for the Red-Dead Sea Conduit project aimed at overcoming the water shortage problem in the region; this is important for economic development, including tourism, housing, commerce, and for the industry of the entire region. However, any future water resources developments must also ensure ecologically sustainable and politically and socially acceptable alternatives.

References

Abdel-Khaleq R. and Dziegielewski, B. 2004. *A National Water Demand Management Policy in Jordan*. Ministry of Water and Irrigation. Amman, Jordan.

Abu-Qdais, H. 2008. Environmental impacts of the mega desalination project: The Red–Dead Sea conveyor. *Desalination* 220, 16–23.

Abu-Qdais, H. and Batayneh, F. 2002. The role of desalination in bridging the water gap in Jordan. *Desalination* 150, 99–106.

Al-Halasah, N. 2003. Water development and planning in Jordan. *Conference on Integrated Water Management Policy Aspects Proceedings*, (Cyprus).

Al-Jayyousi, O. 2001. Capacity building for desalination in Jordan: Necessary conditions for sustainable water management, *Desalination*, 141, 169–179.

Assaf, K., Al Khatib, N., Kally, E. and. Shuval, H. 1993. *A Proposal for the Development of a Regional Water Master Plan*. Israel/Palestine Center for Research and Information IPCRI, Jerusalem.

Beaumont, P. 2002. Water policies for the Middle East in the 21st century: the new economic realities. *International Journal of Water Resources Development* 18, 315–334.

Berkoff, J.A. 1994. *Strategy for Managing Water in the Middle East and North Africa. Directions in Development.* The World Bank, Washington DC.

Berman, L. and Wihbey, P. 1991. The new water politics of the Middle East, in *Strategic Review, Research and Analysis.*

Beyth, M. 2007. The Red Sea and the Mediterranean-Dead Sea canal project. *Desalination*, 214, 365–372.

Brook, D. 1996. Between the great rivers: water in the heart of the Middle East, in *Water Management in Africa and the Middle East* edited by Eglal Rached, Eva Rathgeber, and David B. Brook. IDRC.

CEDARE. 2007. Status of water millennium development goals achievement in the Arab region: Center for Environment and Development in the Arab region and Europe. *Arab Water Council.*

Darwish, A. 1994. The next major conflict in the Middle East water wars – Lecture given at the *Geneva Conference on Environment and Quality of Life*, June 1994.

Elmussa, S. 1993. Dividing the common Israeli-Palestinian waters: an international water law approach. *Journal of Palestinian Studies*, 22, 57–77.

Fisher M.F. and Huber-Lee, A. 2005. Liquid Assets. an economic approach for water management and conflict resolution in the Middle East and beyond, in *Resources for the Future, Issues in Water Resource Policy* edited by Amir, I., Arlosoroff, S., Eckstein, Z., Haddadin, M.J., Jarrar A.M., Jayyousi A.F., Shamir U. and Wesseling H., Washington, DC.

Glueckstern, P. and Fishelson, G. 1992. *Water Desalination and the Red Sea-Dead Sea Canal.* University of Tel Aviv, Tel Aviv.

Haddadin, M. 2001. *Diplomacy on the Jordan: International Conflict and Negotiated Resolution*. Norwell, MA: Kluwer Academic Publishers.

Haddadin, M. and Shteiwi, M. 2004. Water resources in Jordan, evolving policies for development, the environment and conflict resolution, in *Resources for the Future, Issues in Water Resources Policy*, edited by Haddadin, M. Washington, DC, 211–235.

Harza JRV Group, 1996. *Red Sea–Dead Sea Canal Project, Draft Pre-feasibility Report, Main Report, Jordan Rift Valley Steering Committee of the Trilateral Economic Committee.*

Hayton, R.D. and Utton, A.E. 1989. Trans-boundary groundwater: the Bellagio draft treaty. *Natural Resources Journal*, 29, 663–722.

Hillel, D. 1994. *Rivers of Eden: the struggle for water and the quest for peace in the Middle East.* Oxford University Press – Oxford, UK.

Homer-Dixon, T.F., Boutwell, J. and Rathjens, G. 1993. Environmental change and violent conflict. *Scientific American*, 268 (2), 38–45.

Izquierdo, F. 2003. Water Shortages and Conflict Economy and Territory, *Sustainable Development: Panorama the Mediterranean Year.*

Jaber, O. and Mohsen, S. 2001. Evaluation of non-conventional water resources supply in Jordan. *Desalination* 136, 83–92.

Kally E. and Fishelson, G. 1993. *Water and Peace; Water Resources and the Arab–Israeli Peace Process*. Westport Connecticut: Praeger.

Klaas, E. 2003. *Potential for Water Wars in the 21st Century* Presentation to College for Seniors Lecture Series, The World Turned Upside Down.

Kliot, N. 1994. *Water Resources and Conflict in the Middle East.* London, New York: Routledge.

Libiszewski, S. 1995. *Water Disputes in the Jordan Basin Region and their Role in the Resolution of the Arab-Israeli Conflict, ENCOP Occasional Paper No. 13.* Center for Security Policy and Conflict Research/ Swiss Peace Foundation, 52–53.

Malkawi, S.H. 2003. *Wastewater Management and Reuse in Jordan*. Ministry of Water and Irrigation. Amman, Jordan. Wastewater Use in Jordan: An Introduction.

Mohsen, M.S. 2007. Water strategies and potential of desalination in Jordan. *Desalination* 203, 27–46.

Mohsen, M.S. and Al-Jayyousi, O.R. 1999. Brackish water desalination: An alternative for water supply enhancement in Jordan. *Desalination* 124, 163–174.

Molle, F. and Berkhoff, J. 2006. *Cities versus Agriculture: Revising Intersectoral Water Transfer, Potential Gains and Conflicts.* Comprehensive Assessment Report 10, International Water Management Institute, Colombo, Sri Lanka.

Moore, D. and Seckle, D. (ed.) 1993. *Water Scarcity in Developing Countries: Reconciling development and environmental protection.* Winrock International Institute for Agricultural Development, Arlington, VA.

Murakami, M. 1995. *Managing Water for Peace in the Middle East: Alternative Strategies*. Tokyo: United Nations University Press.

Murakami, M. and Musiake, K. 1994. The Jordan River and the Litani. In: A.K. Biswas, (ed.) International Waters of the Middle East. From Euphrates – Tigris to Nile. Water Resources Management Series, No. 2. Oxford: Oxford University Press.

MWI: Ministry of Water and Irrigation. 2002. *Water Sector Planning and Associated Investment Program (2002–2011)*, Ministry of Water and Irrigation, Amman, Jordan.

Naff, T. and Matson, R.C. 1984. *Water in the Middle East: Conflict or Cooperation?* Boulder, Colo., USA, and London: Westview Press. 1–62.

Nazzal, Y.K., Mansour, M., Al-Najjar, M. and McCornick, P. 2000. *Wastewater Reuse Law and Standards in the Kingdom of Jordan*: Ministry of Water and Irrigation Amman, Jordan.

Nortcliff, S., Carr, G., Potter, R. and Darmame, K. 2008. Jordan Water Resources: Challenges for the Future: *Geographical Paper*, 185.

Nsheiwat, Z. 2007. Wastewater use in Jordan: in M.K. Zaidi (ed.), *Wastewater Reuse – Risk Assessment, Decision-Making and Environmental Security*, 73–79. Springer.

NWMP: *National Water Master Plan of Jordan. 1977.* Bundesanstalt fuer Geowissenschaften und Rohstoffe. Hannover/Germany und Agrar – und Hydrotechnik/Essen/Germany.

Potter, R.B., Darmame, K., Barham, N., and Nortcliff, S. 2007. An Introduction to the Urban Geography of Amman, Jordan. *Reading Geographical Papers*, 182.

Salameh, E. 2008. Over-exploitation of groundwater resources and their environmental and socio-economic implications: the case of Jordan, *Water International*, 33 (1) 55–68.

Salameh, E., and Bannayan H. 1993. *Water Resources of Jordan: Present Status and Future Potentials*. Friedrich Ebert Stiftung Amman, Jordan.

Schiffler, M. 1994. Water demand management in an arid country: The case of Jordan with special reference to industry. *Reports and Working Papers of the German Development Institute*. 10, Berlin.

Schiffler, M. 1993. Nachhaltige Wassernutzung in Jordanien. Determinanten, Handlungsfelder und Beiträge der Entwicklungszusammenarbeit. Berichte und Gutachten des DIE No. 7. *Deutsches Institut für Entwicklungspolitik*, Berlin.

Shuval, H. 2000. A proposal for an equitable resolution to the conflicts between the Israelis and the Palestinians over the shared water resources of the Mountain Aquifer. *Arab Studies Quarterly* 22 (2), 33–61.

Shuval, H. 1992. Approaches to resolving the water conflicts between Israel and there neighbors-regional water for peace plan. *Water International*, 17, 133–143.

THKJ, MWI, GTZ. 2004. *Water Sector Planning Support Project (WSPSP).* Amman, Jordan: Ministry of Water and Irrigation.

WAJ: Water Authority of Jordan. 2005. *Open files.* Amman, Jordan.

Wolf, A.T. 1995. *Hydro-politics along the Jordan River: Scarce Water and its Impact on the Arab-Israeli Conflict.* Tokyo:United Nations University Press.

Yakhin, Y. and Clayton, W. 2006. *Water in the Israeli-Palestinian,* James A. Baker III Institute for Public Policy of Rice University.

Abbreviations and Acronyms

GNP	Gross Net Production
JICA	Japan international cooperation agency
L	Liter
MCM/yr	Million cubic meters per year
NWMP	National water master plan
ppm	Parts per million
RDSC	Red-Dead-Sea Conduit
TDS	Total dissolved salts
WAJ	Water Authority of Jordan
m^3/yr	Cubic meters per year

mg/l	Milligrams per liter
JRV	Jordan river valley
m^3	Cubic meter

Chapter 8
Water Loss Management in Cyprus[1]

Symeon Christodoulou

Introduction

Each year, hundreds of kilometers of pipes across the globe need upgrading or replacing in an attempt to maintain the uninterrupted transport of water. The problem of aging infrastructure and of associated water losses in urban water distribution networks has been one of the biggest infrastructure problems facing city and municipal authorities and a major task in their efforts to achieve efficient and sustainable management of water resources. Interestingly enough, as reported by the International Water Association (IWA), even in developed countries the "unaccounted-for" water is in the range of 20 percent to 30 percent, whereas in developing countries this percentage is even higher. According to studies found in the literature for example, water losses in France's water distribution network have been estimated at an average 26 percent, in England/Wales at 19 percent and in Italy 29 percent (MCG 2006). Yet local communities have poor prediction tools to prioritize how essential infrastructure investment should be conducted.

The driving forces behind pipe-replacement capital improvement projects have primarily been the mandate to safeguard the health of urban populations, the need to increase the reliability of the pipe networks and the service provided to people, as well as socioeconomic factors related to the cost of operations and maintenance of water distribution networks.

The accidental or deterioration-based breakage of water distribution systems represents a range of health and safety problems, which may unintentionally place the health of thousands of citizens at risk during even a single break. Existing water distribution networks are increasingly at risk due to numerous factors (both internal and external to the distribution networks), yet important advances in the development of expert systems for utility system risk assessment have been confined to immediate, pipe-related characteristics (e.g. age, geometry, operating pressure, leakage history) (US Environmental Protection Agency 2006) while

---

1 The information included in this chapter constitutes a summary of the data obtained through the generous funding of USA's National Science Foundation (NSF 2001–2003) and the Cyprus Research Promotion Foundation (CyRPF 2005–2008) and provides an overview of proposed water management actions and the benefits expected to stem from them.

recent advances in remote monitoring have yet to be embraced by the water utility industry.

Sustainable Management of Water Distribution Networks

Water scarcity and the added threat posed by the altered climatic terms have intensified the need for the development of suitable approaches for improved management of aquatic reserves. Such methods should aim not only at saving water but also at safeguarding the balance between water supply and water demand. Within such a framework and especially in arid countries, minimization of water losses in urban networks should be among the primary objectives not only of the network owners/administrators but also of individual consumers. It is, consequently, imperative that the administrators of water supply (and irrigation) networks apply effective (preferably simple) methodologies for the reduction of water losses that can stand the test of time.

Sustainable management of urban water distribution networks should include not only new methodologies for monitoring, repairing or replacing aging infrastructure, but more importantly it should facilitate the modeling of the state of each piping network and the predictive analysis of its behavior over time. The goal should be the assessment of deteriorating piping infrastructure and the utilization of historical incident data and risk-of-failure metrics for devising intelligent "replace or repair" strategies.

The term "replace or repair dilemma" refers to the decision made by water-distribution network administrators at various points in time about whether to replace a failing pipe based on their evaluation of the situation at hand. Exercising the replacement option would result in incurring replacement costs at the time of action but it would lower the risk of failure in the future. Exercising the repair option would mean retaining the pipe in consideration and thus the saving of any replacement costs, but it would result in a higher risk of future failures and thus in increased repair costs, disruption of service and damage to neighboring facilities. In essence, the decision to repair or replace failing pipe segments boils down to a water agency's tolerance of risk, its appraisal of the situation and the associated risk level, and to the socioeconomic implications from a possible pipe failure.

The "replace or repair" dilemma facing water utility agencies should thus be at the core of any effective and sustainable effort to better manage piping networks, since successful resolution of this problem would help in reducing water losses and life-cycle costs as well as in increasing the reliability of a network and the quality of its service to citizens. In fact, life-cycle costing and maintenance strategies become of paramount importance to water agencies as they seek ways to increase system reliability and quality of service while minimizing operational costs. At the outset of this balancing act of operating costs and reliability lie two important issues: (i) should an organization repair or replace ageing and/or deteriorating water mains and, (ii) in either case, what should the sequence of any such repairs be as part of a long-term network rehabilitation strategy?

With that in mind, water distribution agencies are nowadays faced with the increasingly more complex task of intelligently and efficiently assessing (or modeling) the condition of a pipe network and managing the network in ways that maximize its reliability and minimize its operational and management costs. Research to date has helped identify a number of potential risk factors that contribute to pipe breaks, such as the pipe's diameter, material and length, the operating pressure in the network, a pipe's age, the number of previously observed breaks for each pipe segment, the soil conditions, and the external loads to the underground piping network. Some of the factors are time-invariant and some are time-dependent, but all have been shown to be contributing factors to the overall risk-of-failure level.

The Eastern Mediterranean Basin and the Middle East: The Case of Cyprus

The extent of the rapidly worsening water scarcity problem and its water-loss parameter is most clearly manifested in southern Europe (Spain, Portugal, Italy, Malta, Greece, Cyprus) and in western Middle-East countries (Israel, Lebanon, Turkey). The first group consists of European Union (EU) countries that, despite being among the poorest economies in the EU, have the means to implement holistic technology-based and EU-financed water management strategies. The second group consists of countries that have access to large quantities of water (Turkey, Lebanon) or the technological and financial means to mitigate the lack of it through desalination plants (Israel). In the past, water-loss management was not among the top national priorities for any of these countries. In the last few years, though, water-loss management and water conservancy have become pressing needs and a national cause for all of these countries.

Among the most notable cases is that of the Republic of Cyprus, a small island of about 800,000 inhabitants located in the eastern corner of the Mediterranean (neighboring Greece, Turkey, Syria, Lebanon, Israel, and Egypt). The mean annual precipitation between 1901 and 1970 was approximately 500 millimeters, whereas during the period between 1971 and 2000 the annual average rainfall dropped to 460 millimeters. The quantity of water, which corresponds to the total surface of the Government controlled area, is 2,670 million cubic metres (Mm^3) but only 370 Mm3 of this (about 14 percent of the total volume) is available for development (the remaining 86 percent returns to the atmosphere through evapotranspiration). The mean annual quantity of 370 Mm^3 of water is distributed between surface and groundwater at a ratio of 1.75:1.

The two highest water-consuming sectors on the island are domestic use and irrigation, with the former amounting to about 20 percent of total water demand and the latter to about 69 percent. The remaining quantities correspond to tourist demand (5 percent), industrial (1 percent) and environmental purposes (5 percent).

As the EU Water Initiative (EUWI) reported in a recent study (EUWI 2006), Cyprus is among a group of six European countries classified as "water stressed." This assessment is based on a metric termed "water exploitation index (WEI)," which is defined as the mean annual total demand for freshwater divided by the long-term average supply of freshwater resources.

Three of the other five countries in this category are also located in Cyprus's neighborhood (Malta, Italy, Spain) and are subject to similar climatic conditions but only Malta has the same physical constraints as Cyprus's.

Cyprus's water shortage problem and worsening water-related situation, even though not completely identified with the change in global climatic conditions, is undoubtedly affected by it. Until recently, the island was highly dependent on the annual volume of rainfall and had a policy of "water storage" rather than "water production" for securing the necessary quantities of water. In fact, Cyprus is among the European countries with the highest percentage of stored water volume (over 20 percent) in relation to their annual renewable freshwater resources (Turkey and Spain being the other countries) (EUWI 2006). These countries also use the highest percentage of their water resources for irrigation (EUWI 2006), an activity that demands the largest water volume in the driest seasons and requires winter storage.

This general policy of water storage led Cyprus's government, at least in the first years following the island's independence in the 1960s, to a policy of dam construction in the hope that there would be not only enough rainfall to meet the island's annual needs but also enough water for storage and even enrichment of the island's underground aquifers. In the years since the 1950s, the total water capacity of the island's dams increased from about 6.2 to 327.0 Mm^3. It is also worth noting that about 265 Mm^3 (roughly 80 percent of the total dam capacity) was added after 1974, the year that Cyprus was forcibly partitioned as a result of a Turkish invasion and military occupation of the northern part of the island. The island has since been divided into the internationally recognized Republic of Cyprus and a breakaway state in the Turkish-occupied north, recognized only by Turkey. Not only did approximately 37 percent of the Republic's territory then come under Turkish occupation, but also about 70 percent of the island's economic potential.

In terms of water resources, the Republic lost access to the island's main groundwater aquifers in the Western (Morphou) and Eastern Mesaoria (Famagusta) plateaus and was forced into hurried and difficult decisions about water supply and demand; this put a tremendous strain on the island's water management policies. The strain resulting from the loss of the aquifers has recently been slightly alleviated by efforts to achieve bi-communal water management policies and projects. For example, a recent grant of 47 million euro by the European Union will help finance the construction of a new sewage treatment plant in the northern part of Nicosia (the capital of Cyprus) to replace the current outdated facility.

The project is scheduled for completion in 2009 and will treat wastewater so that it can be recycled for irrigation on both sides of the island. After all, despite the forced division of the island, both Greek and Turkish Cypriot communities share a common ecosystem which recognizes no boundaries and requires bi-communal cooperation to ensure its preservation. Other bi-communal projects include the extension and modernization of the sewage piping network in Nicosia, joint workshops on sustainable water management policies, and joint management and assessment of the ecology of Cyprus's artificial wetlands.

Unfortunately, despite these recent efforts to mitigate the effects of water scarcity, the problem has been increasingly compounded by the worsening climatic conditions and the increase in water consumption. The deteriorating climatic conditions, the observed greenhouse and desertification effects, the decreasing annual rainfall and the increase in the pace of urbanization and incoming tourism have overloaded and negatively impacted the island's overall water situation. For five consecutive years (2003–2008) the total rainfall on the island was in decline and below the expected annual average and dams were close to drying up. By the end of summer, 2008, water in storage occupied only 5 percent of the total dam capacity (about 15.0 out of 273.0 million cubic meters of water). To make things worse, the total volume of water required for satisfying the island's annual needs reached 75.0 million cubic meters per annum. The underground aquifers have been overdrawn by about 40 percent per year and are slowly being destroyed by the intrusion of saltwater. The dismal image that the dams on the island currently present proves that water storage alone is not a prudent policy.

In addition to exogenous factors (climatic conditions, desertification, urbanization), several endogenous factors have also added to the worsening of the island's water problem. The most important of these factors is the water loss in urban distribution networks. A recent study of the situation by the Republic's Auditor General showed a "water loss" of about 8.0 million cubic meters, in a year during which the total flow of water in the dams was roughly 27.0 million. The assessment that the losses correspond to roughly one third of the total water stored in the dams can only cause intense concern. Added to this is the equally worrying assessment of the Auditor General that in municipalities and communities other than the three major ones (the cities of Nicosia, Limassol and Larnaca) the loss of water exceeds the 30 percent mark. For the island's three major urban centers, water loss is estimated at about 7.5 million cubic meters per annum (roughly 18 percent of the total volume circulated in the network).

These losses are diachronic and the problem is continuously compounded because of the aging of distribution networks and of the deterioration of pipes. The problem has become even greater because of the difficulty in routinely examining the state of piping networks and the inability to forecast the time and location of future failures.

Asset Management and Strategic Planning

The International Water Association (IWA) has recently established a method for "water accounting" which records and checks the quantities of water produced within a region of interest and then monitors these quantities as they are trafficked in the piping networks and consumed by consumers. In the end, the networks' operational efficiency is expressed in metrics that are internationally comparable. The metrics express "revenue" and "non-revenue" water and relate these quantities to other factors that can help the network administrators identify possible problems and thus mitigate water losses.

Non-revenue water (NRW) is water that has been produced but "lost" before it reaches consumers. Losses can be real losses (through leaks) or apparent losses (for example through theft or metering inaccuracies). NRW is typically expressed as a percentage of the total volume of water produced, or as the volume of water "lost" per km of water distribution network per day. High levels of NRW are detrimental to the financial viability of water utilities as well to the quality of water itself. Apparent losses are typically in the order of 2–3 percent and real losses should be kept to as small a percentage as possible.

A comprehensive and successful water-loss management policy should thus aspire to the following:

- reduction of water losses and implementation of proper water auditing policies,
- creation of mechanisms for the prioritization of pipe repair or replacement strategies in the most effective and financially viable way,
- strengthening of water-related citizen awareness,
- longevity and viability of the policies that will be applied,
- coupling of research and practice for improvements in the already-applied strategies,
- technological growth and adaptation of results of relevant studies, knowledge transfer and adaptation to the needs of local societies.

The Water Board of Limassol recognized at a very early stage the importance and significance of establishing a proper water audit system according to IWA's guidelines and has, over the years, developed its infrastructure in such a way as to be able to account efficiently and accurately for all water produced or "lost" (non-revenue water) in the city's water network. Reduction and control of water loss was achieved through the application of a holistic strategy based on the approach developed by IWA's Water Loss Task Force. An integral part of this approach is the establishment of a strategy for pipe break incidents. The policy further envisions the prioritization of the repair/replacement actions on the basis of risk of failure, life-cycle costing, social and financial impacts.

Following a strategy of redesigning its network into well-defined equi-pressure water zones termed "District Metered Areas," the Water Board also developed mechanisms for evaluating historical incident data as to when and how water pipes break. The goal was twofold: (i) to minimize water losses by lowering the pressure and its fluctuation in pipes, and (ii) to identify possible data patterns in the behavior of the network and utilize these patterns in predicting new breaks. The analysis of historical data was based on a number of analytical and numerical tools and it investigated the possible contribution of a number of presumed risk factors to a "break or not" outcome. Furthermore, it allowed the ranking of these factors according to their relative importance and contribution to the predicted "break or not" output and the pipe's lifetime.

The analysis of whether to repair or replace a burst pipe, and thus the Board's asset management strategy was finally tabulated in a database management system and then mapped on a geographical information system (GIS) that enables users to view results graphically and query them spatially (on a city map).

Initially, historical data on previously observed breaks (NOPB) are lumped at a street level and then mapped to a GIS map of the pipe network, and color-coded to indicate the variable degrees of their inherent risk of failure (Fig. 8.1). Moreover these incidents are categorized by reason of failure and also depicted graphically on the GIS map. The Agencies can therefore easily and holistically review the status of their network in terms of where and how often pipes break, view the computed risk-of-failure and the expected lifetime for each segment of the pipe network and thus be provided with intelligent DSS tools for deciding on when to replace pipes so as to minimize the risk of failure and water loss.

Decision Criteria

Decision criteria can be obtained from the knowledge hidden in the historical data. Various modeling tools permit the identification of patterns in the interactions of the underlying risk factors and their possible contribution to a pipe's failure. In the first stage of a two-stage analysis all historical data are processed to identify the most important risk factors and rank them according to their relative importance and contribution to pipe failure (Fig. 8.2). The analysis can provide the Water Board with an insight as to the survivability of each pipe category over time. For example, if the data is grouped by the break-incident type then a "hazard plot" can be produced (Fig. 8.3). Such a plot reveals that, for the data under study, the hazard rate related to pipe deterioration greatly outpaces the hazard rate of the other incident types (corrosion, interference by others, tree roots, connection hose). What starts as a relatively small deviation at the beginning of the observed failure time (at about 30 years) soon doubles (values of 0.008 vs. 0.004) at the upper end of the expected lifetime (at about 35 years). And while the hazard rate for incidents attributed to tree roots, corrosion, connection hose or other causes increases almost uniformly over time, the deterioration-related hazard rate accelerates over time (from 0.001 to 0.008). This is an indication that aging pipes should be replaced at smaller time-intervals (at about 11,000 days, or 30 years); otherwise the risk of

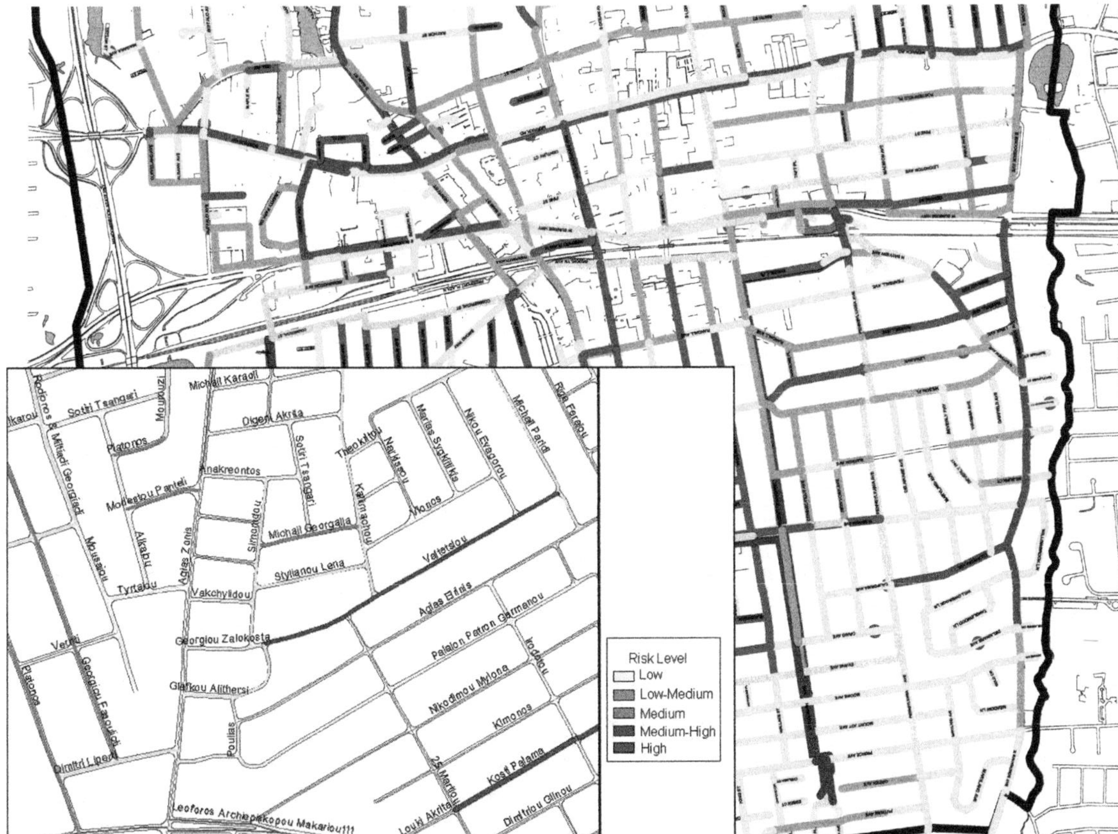

Figure 8.1 GIS mapping of risk-of-failure

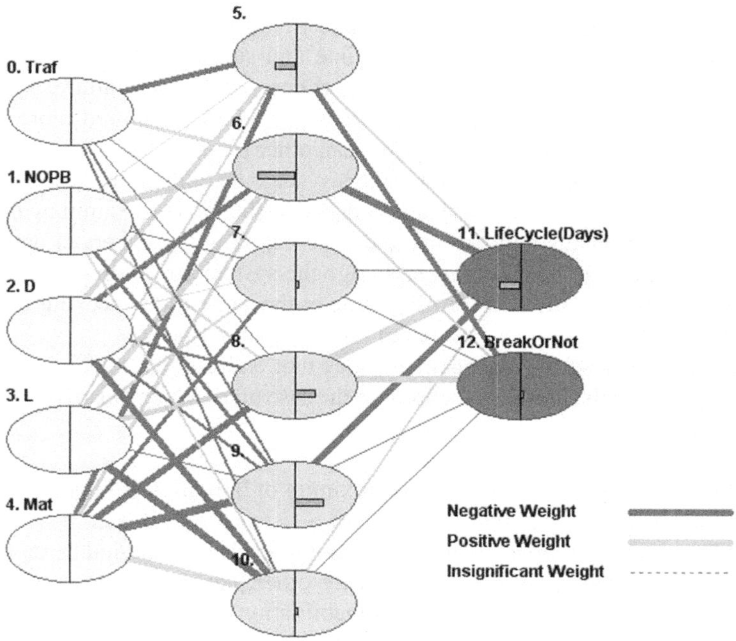

Figure 8.2 Artificial neural network (ANN) implementation of the risk-of-failure analysis of water mains

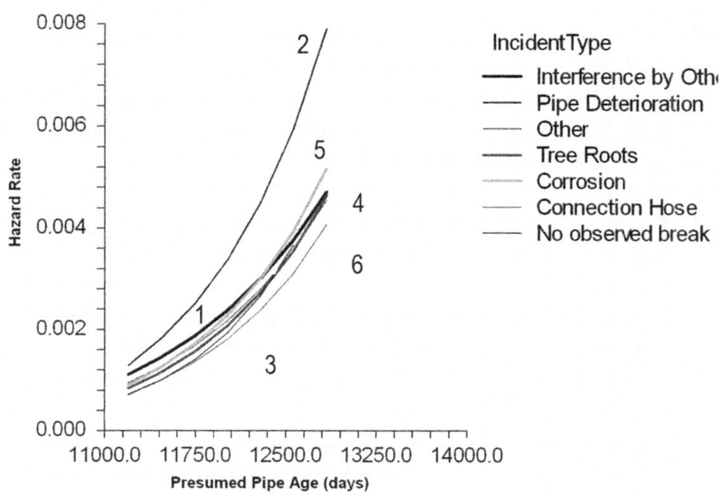

Figure 8.3 Hazard rate vs. presumed pipe age (in days) for data stratification based on incident type

failure increases substantially. Also of interest is the observation that the hazard rate related to tree roots accelerates in time and surpasses the rate of increase of corrosion-related, interference-related and hose-connection failure. This is an indication that pipes in the vicinity of trees should be monitored more closely and/or replaced at smaller time intervals than other pipes.

This is followed by a second analysis stage in which the knowledge is transformed into rules. The knowledge deduced is obtained by examining different combinations of inputs and outputs and by combining the observed patterns in behavior with expert rules. In this last stage the system builds on the knowledge acquired and transforms it into rules, while at the same time checking the rules' validity and conformity to the observed input/output data sets.

The result is a set of management rules that address the "repair or replace" dilemma posed in the previous sections. In the case of the Water Board of Limassol, the top-tier rules are as follows:

- Priority is given to areas in the proximity of buildings of high public value (e.g. hospitals, schools).
- Priority is then given to areas combining residential and industrial use.
- Priority is then given to areas where other planned construction work is taking place (such as roadway rehabilitation) so as to maximize parallel work and minimize successive disruption.
- Priority is then given to pipes with a high number of observed previous breaks (NOPB).
- Then to pipes with large diameter take precedence.
- These are followed by pipes made out of cast-iron, followed by steel pipes and plastic pipes.
- Finally, pipes subject to heavy traffic loads take precedence.

Each of these rules is subdivided into second-tier rules, as needed. For example, when considering the effects of previous breaks, diameter, and material, the following rules were also developed:

- MDPE-Black pipes should be replaced approximately every 31 years, galvanized and asbestos-cement pipes every 33 years, and MDPE-Blue pipes every 35 years.
- Since pipe deterioration is the risk factor that accelerates the most over time, pipes should be checked regularly (especially when they approach their expected lifespan) and replaced when an acceptable hazard rate threshold is reached.
- Tree roots are a risk factor whose hazard rate accelerates in time. Special care should be given in cases where pipes are laid in proximity to trees and precautionary measures should be taken.

Conclusions – Recommendations

Easy or magical solutions for solving a country's water scarcity problems do not exist. On the contrary, most mitigation measures require meticulous planning, significant investments of time and money, and the combined efforts of the public and private sector, research institutions, and citizens. Many of the proposed measures have been highlighted in the past. Most notable of these measures, especially for countries with arid climates, are:

- *Autonomy from climatic conditions – Use of desalinated and/or recycled water*: In light of the worsening global climate conditions, a society's independence from climate fluctuations should be a primary goal. Much has been done but even more can be achieved. Countries with arid weather conditions should aim at a strategic transition from a policy of storing water (i.e. dams) to a more active policy of generating water (i.e. desalination or sewage treatment plants). This policy shift should be twofold, with a parallel increase in the use of recycled (grey) water and an increase in the volume of desalinated water introduced into the national water balance sheet so as to decrease our dependence on rainfall.
- "*Grey water*": In order to reduce the water deficit, the Government of Cyprus has adopted novel measures for water conservation using "grey water." Among them are the provision of subsidies to households for saving good-quality domestic water and instead using the water from private boreholes on their property for toilets (through the connection of toilet tanks with a secondary piping network to the boreholes), or subsidies for the installation of "grey-water" recycling systems in houses and public buildings and the use of the grey water for watering gardens and toilet flushing. The secondary piping system allows for lightly polluted or "grey water" from baths, showers, hand or wash-basins and washing machines to be kept separate from heavily polluted water from WC's and kitchens. As a result, it is relatively easy to intercept each type of wastewater at the household level for subsequent treatment and re-use. With this scheme, Cypriots have achieved a drinking water conservation of 30 percent to 65 percent (EUWI 2006).
- *Enrichment of underground aquifers and water over-pumping*: The drilling of private boreholes on each household property or for irrigation purposes comes at a price. Over-pumping of water deprives countries of precious quantities of water, puts a strain on strategic water reserves that are not easily replenished, and deprives future generations of this precious natural resource. Over-pumping should therefore be avoided through the implementation of strict national policies. In addition to drilling and pumping laws to regulate the number of boreholes allowed in each area and the volume of water each one may collect, national governments should

also implement measures for the enrichment of underground waters by use of "grey water" (i.e. water produced through wastewater treatment).

- *Cultivation of the public's awareness and sensitivity to water-related issues*: Even though the recent drought in Cyprus (2005–2008) and the measures taken to address the situation have caused a variety of socio-economic problems affecting the general population, several lessons have been learned from the situation. For example, among the measures taken were water supply restrictions (about 10 hours of water supply in every 48 hours), demand management measures, and supply enhancement measures. Nevertheless, the problem of continuous cultivation and intensification of public awareness and sensitivity to water-related issues is diachronic and it should be a top priority for the agencies involved. After all, increasing water supply is not enough if not paralleled by the curbing of water demand. Some Mediterranean countries with arid climates have already made progress in raising the consciousness of their citizens about water-related issues. In the framework of such policies, for example, citizens are given subsidies towards the (1) installation of domestic water-recycling equipment for the reuse of grey water in the lavatories or for gardening purposes; (2) drilling for water in their household plots and connecting of the borehole with the domestic piping network for the use of the water in the lavatories. This policy of incentives should be combined with a policy of disincentives, through the policing and imposition of fines whenever applicable so as to curb wastage in water usage.
- *Water loss reduction and sustainable management of urban water distribution networks*: This parameter is of the utmost importance to all national and regional governments particularly to countries with arid climates. Only through the development and implementation of holistic and systematic "repair or replace" asset management strategies can it be achieved.
- *Water balance*: IWA guidelines on water balance and water accounting should be adopted immediately by all agencies. Only through proper water accounting can the problem of non-revenue water that lurks in most urban centers be identified and properly addressed.
- *Reorganization of water distribution networks*: Since pressure in water pipes is a significant risk-of-failure parameter, water distribution networks should be reorganized in ways that minimize pressure values and the pressure fluctuations in them over time. To that effect, IWA has already suggested the use of urban network zones having approximately equal pressure levels and similar function and usage over time.
- *Rehabilitation of ageing water distribution networks*: Water Boards and other regional agencies, with the financial support of national governments, need to invest in the renewal of aging water piping networks and in long-term and sustainable network management technologies. The policies to be implemented should be based on a structured approach and in

accordance with scientific methods that take into consideration the deduced risk of failure and expected lifecycle of each pipe segment. The ad-hoc pipe replacement that is so often employed by agencies should immediately be abandoned. Reactive measures should be replaced with proactive ones.

- *Performance metrics and financing schemes*: At the same time, states should apply techno-economic methods for the evaluation and financing of water agencies (or their policy with regard to water pricing) by use of properly devised performance indicators. Such performance indicators should relate financing to the performance of the networks in terms of water loss and service downtime. Any policy adjustment and water pricing differentiation (e.g. per community, per type of use, per volume of consumption etc.) should adhere to the performance metrics and performance goals set nationally.
- *Water pricing*: Among the most powerful tools that water agencies have in curbing water demand is water pricing. In fact, EU regulations call for water prices that fully reflect the cost of producing and delivering water to consumers. This means that water cannot be subsidized by national or regional governments and that, despite being a basic social good, water should be priced appropriately. The questions then become "How cheap is expensive enough?" and "How can we arrive at a price that reflects water's true value?" The answer to either question is difficult to arrive at, and many studies have attempted to address them. Both economic and social factors can and should be taken into consideration. In the end, the evaluation and application of a just and equitable water pricing policy can give citizens both incentives and disincentives and aid appropriate water consumption. Countries like Cyprus, for example, are on the verge of implementing a combination of stepwise pricing policies that resemble income tax policies (different rates apply to different water consumption ranges) and water quotas per household. Another possible measure is the implementation of "water markets" much like those of carbon emission, in which citizens may opt to purchase water to add to their household water quota from citizens not in need of their allocated water quantities.
- *Organizational reform*: The Water Boards (especially the Technical Services Divisions within the organizations) should immediately recognize the usefulness of scientific analysis and relevant technologies. Their corporate memory and knowledge of personnel usually prove to be their most valuable assets. On the other hand, the lack of expertise in the new technology is proving to be the biggest obstacle to widespread adoption of new technologies. What is needed is the means to embrace scientific progress and to develop and manage technological change across a corporate structure in an effective and sustainable manner that takes into consideration not only the advancement of science but also the status of the organization's existing personnel and modus-operandi. The technology can be implemented gradually and in phases. Organizational reforms should also include modernization, automation and integration within the

organizations, continuous training of their personnel, collaboration with research institutions, application of state-of-the-art technologies, and frequent and institutionalized exchange of knowledge between the various stakeholders.

- *Reorganization of agriculture and proper choice of suitable cultivation*: It is generally accepted that the agricultural sector of a country's economy greatly impacts its demand for water. The problem is especially complicated in third-world countries that rely heavily on agriculture and in countries with arid climates. Such countries need a general revision of their agriculture sector so that water scarcity issues are also factored into the choice of plant varieties grown in each region. Water-intense varieties should be replaced with more water-efficient crops. The issue of irrigation and suitable crops is also closely linked to the issue of water pricing. In the case of Cyprus, water for irrigation is supplied through government and non-government schemes. In the former case water is delivered directly to individual farmers (retail supplies) and in isolated cases may also be provided on a bulk basis to irrigated sectors. In this case, water prices are established on a volumetric basis and are uniform for all schemes, covering a high proportion of the total financial cost. This, in effect, can be considered a government subsidy and until recently, it was not tied to the type of cultivation, the necessity for it and the resulting product quantity, rendering the whole scheme inefficient and wasteful of water. Non-government schemes consist of smaller irrigation schemes which are managed by committees chaired by the District Officer. Prices may vary, depending on the location, volume and cultivation for which water is supplied. This scheme is more likely to put pressure on producers to curb water demand based on the suitability of the crops chosen.
- *Unified national water resources management entity*: Last, but not least, and for all the above reasons, it is imperative that a unified water resources management entity be established at the national level to devise and implement strategies and policies pertaining to water.

References

Aslani, P. 2003. *Hazard Rate Modeling and Risk Analysis of Water Mains*. MSc. thesis, Polytechnic University, Cyprus.

Charalambous, B. 2005. *Experiences in DMA Redesign at the Water Board of Lemesos, Cyprus*. International Water Association's Specialised Conference on Leakage, Halifax, Nova Scotia, Canada. 403–413.

EUWI 2006. *Water Scarcity Management in the Context of WFD*. MED Joint Process WFD/EUWI, Water scarcity drafting group (WGB/15160506/25d).

Metropolitan Consulting Group (2006). VEWA survey – Comparison of European water and wastewater prices.

US Environmental Protection Agency. 2002. Decision-Support Tools for Predicting the Performance of Water Distribution and Wastewater Collection Systems. EPA/600/R-02/029. Available at http://www.epa.gov/nrmrl/pubs/600r02029/600r02029.htm.

Chapter 9
Water and Geology in the Mediterranean

George Stournaras

The Mediterranean Basin, including the Black Sea Basin region, comprises 24 countries with some of the most heterogeneous eco-climatic topographies whose groundwater quality is now under threat due to human activities that are rapidly affecting the region's hydrological cycle. Groundwater has always been important to the region. According to Andreo and Dura (2008) cuneiform tablets provided the first reference to hydrological exploration, with an expedition in 852 B.C. by Assyrian King Salmanassar III to the headwaters of the Tigris, where spring water was obtained to supply his city. In Biblical times, groundwater was used for water supply to Sinai, Jerusalem and other cities.

Although the philosophers of Ancient Greece already developed some basic concepts used currently in hydrogeology, such as the hydrologic cycle, water sources, and the quantity and quality of water, understanding the proper management of the aquifer systems in the Mediterranean basin is far from complete. Complicating matters even more, the states that comprise the Mediterranean region contain significant differences in social demographic factors such as water usage customs, level of economic development, organization of agriculture and agri-business, patterns of urbanization and types of human settlement.

Historically, the Mediterranean Basin has been affected by climatic changes in terms of its water regime due to its geographical position and to the circulation of wet and dry air masses.

The prevailing arid and semiarid climatic conditions as well as periodic droughts have resulted water-related problems in Mediterranean countries since the 19th century. Population is constantly increasing and tourism is expanding, putting an ever increasing pressure on the existing aquifers system, and contributing to the deteriorating situation.

In this chapter we will first discuss the similarities and differences in the aquifers and groundwater supply. Then we will consider the impact of human activity and climate change on the availability and quality of groundwater, now and in the future

Similarities and Differences

Mediterranean countries present many common features and, at the same time, a considerable variability in terms of climate, water and land resources as well as development issues. These features include arid and semi-arid climates, limited water resources, agricultural development related to water availability, and the high economic and social value of water. From a geological and hydro-geological point of view the common features include:

- In the Mediterranean main aquifers are developed in alluvial deposits, (porous aquifers).
- Additional aquifers are in carbonate rocks (karstic aquifers).
- Finally, the extended igneous and metamorphic rocks, non karstified carbonate and hard sedimentary rocks form an important field of fissured rock aquifers (discontinuous media).
- Common to all coastal aquifers is the problem of the salinization, mainly due to over-exploitation resulting from intense social, economic, industrial and agricultural activity occurring over a period of more than 25 centuries.
- In addition, the Mediterranean basin is the world's foremost tourist destination resulting in exhaustion of the scarce water resources on many of the Islands.

On the other hand, the differences among the Mediterranean countries' hydrologic regimes and water supply include:

- The difference in the geological structure between the northern and eastern coasts and the southern coast of the basin;
- The circulation of the aerial masses, contributing to the climate and precipitation regime formation;
- The difference of alimentation of the basin with fresh water between the northern coast (the Ebro and other Spanish rivers, the Arno, Po, Rhone, and Tiber, rivers of the eastern Adriatic coast, Vardar and other Greek rivers) and the southern coast (mainly the Nile River). These rivers also transport the pollution of human and other activities in the interior of the Mediterranean countries to the sea;
- The special character of the Dinaric karst on the East Adriatic coast;
- The remarkable difference between the north and south coast of the basin, namely the dry and desert areas of the African coast.

Human-induced Changes in the Aquifer Systems

In order to understand the human effect on the aquifer system it is important to consider the hydrological cycle, because as shown in chapter six, changes in water use at the soil surface directly affect the amount of ground water that can be withdrawn safely. The hydrological involves the continuous circulation of water in the Earth-atmosphere system. Water is transferred from the seas and oceans through the atmosphere to the continents and back to the oceans. As the precipitation reaches the earth's surface, this precipitation is divided into either evapotranspiration, run-off and infiltration. Run-off in turn is subject to secondary evaporation and infiltration and the water already infiltrated which emerges through the discharge of springs and is added to the run-off. Furthermore, water that infiltrates into the soil will be retained in the subsoil and used by by the plants where the main root systems exist. Water in excess of the plants' needs is subject to a percolation that recharges the groundwater tables. At all stages of the procedure outlined above, a chemical reaction between the water and the geological environment takes place, affecting both the water and the soil. Hence, it is obvious that every disturbance in the precipitation regime (quantity, quality, rhythm, duration, intensity etc.) effects important modifications in the hydrologic (surface water) and hydrogeological (groundwater) balance.

The Hydrological Cycle and Geography

The beginning of the phenomena of the hydrologic cycle is precipitation (rain, snow, hail). As it reaches the earth's surface, this precipitation is subject to a protogenic division into evapotranspiration, run-off and infiltration. There is also a secondary division concerning the water of the run-off, which is subject to secondary evaporation and infiltration and the water already infiltrated which emerges through the discharge of springs and is added to the run-off. Furthermore, the infiltrated water gives priority to covering the necessities of the soil and subsoil zone (retention, capillary, absorption water) and the hyporheic circulation, where the main root systems exist. Each time the eventual surplus of the infiltrated water is subject to a percolation that feeds the groundwater tables. At all stages of the procedure outlined above, a chemical reaction between the water and the geological environment is accomplished, effecting modifications on both sides. Hence, it is obvious that every disturbance in the precipitation regime (quantity, quality, rhythm, duration, intensity etc.) effects important modifications in the hydrologic (surface water) and hydrogeological (groundwater) balance.

One of the main problems of using more water than is being recharged to the groundwater is the salinization of the coastal aquifers in the Mediterranean basin. Excess water consumption also leads to waste production, which percolates down to the groundwater, polluting the aquifer. Moreover, the principal rivers of the region discharge into the Mediterranean Sea, adding both an additional recharge and an additional load of pollution originating from the continental activities.

Based on the discussion above, there are four types of aquifer each of which is impacted differently by human use:

- Unconfined aquifers, of high permeability, rare in coastal areas,
- Unconfined aquifers of medium to high permeability, under exploitation and salt water intrusion conditions,
- Unconfined to semi-confined aquifers of high permeability and moderate exploitation and sea water intrusion conditions,
- Confined aquifers, of low to moderate permeability, underexploited, without salt water intrusion problems.

Table 9.1 (adapted from Deer and Patton, 1971) lists the various mechanisms by which the aquifers are impacted by the different processes that affect the fate of contaminants in the environment. Continuous media in general are able to reduce the concentration of the contaminants to a greater extent than other media. Microbiological degradation can be an important mechanism for reducing the concentration. However some compounds are so toxic that microbial degradation is not effective.

From a hydrogeological point of view, most of the coastal aquifers are in karstic and fractured rocks, which makes them extremely vulnerable to salinizationm and pollution by surface sources. The reasons why they are so vulnerable are as follows:

- They are open at the surface.
- Karst formations transport pollution more rapidly than porous media.
- Removal of pollution due to absorption and retentione through soil is absent.

The resource contamination and the resource contamination are two distinct processes. There is a possibility of direct infiltration of karstic aquifers by a bypass of the over-contamination of strata through karstic geoforms, such as sinkholes etc. Table 9.2 is adapted from the pioneering work of Deer and Patton (1971).

Deltaic formations are currently near depletion due to the construction of the dams in the upstream areas and of embankments within the deltaic area. Because

Table 9.1 Comparative effectiveness of porous media in reducing contaminant concentration

	Mechanism					
Media ↓	**Dilution**	**Infiltration**	**Absorption**	**Retaining**	**Chemical reactions**	**Microbiological degradation**
Continuous media	+	++++	++++	++++	++++	++++
Discontinuous media	++	+++	++	+	++++	++++
Karstic media	+++	+	-	-	++	++++
Surface media	++++	-	-	-	+	++++

of the lowering of the water table in the aquifer below sea level, salt water flows inland, making all the fresh water unsuitable for drinking.

Finally, climate change will cause an increase of the mean global temperature of 0.9°C for the sea environment and of 1°C for the land. Most of the global climate models predict that in the near future there will be rapid fluctuations, causing periods of extreme heat followed by a decrease in cold periods. Precipitation will be reduced because of the frequency of hot dry periods and evaporation will increase. The essential effects will be the following:

- Reduction of the water availability.
- Increased evaporation.
- Increase in the demand for water for irrigation.
- Increase in the demand for public water supply and general use combined with an increase in population exacerbated by the predicted increase in longevity.
- Increase of the salinization of the coastal aquifers.
- Increase of the concentration of contaminants within the fresh water bodies due to the increase of pollution, reduced water accumulations and increased withdrawal.
- Increased use of air-conditioning units.
- Reduction of the operation of water power stations due to reduced available water.
- Problems in ecosystems and wetlands.

All of these effects should be considered as a common heritage confronting the Mediterranean Basin. In addition to these serious threats there other anticipated effects for the basin that should be stressed:

- Reduced replenishment of the sea water, mainly from the Straits of Gibraltar and secondly from the Suez Canal;
- Contamination originating from human activities in coastal areas and additional contamination from these activities in continental areas arriving at the shoreline through the flow and discharge of rivers;
- Variation concerning the organization, degree of progress, and implementation of the Water Frame Directive among the EU countries and the states of Africa, Middle East and the Black Sea;
- The same differentiation among the EU countries;
- Day to day functioning and the citizens' participation focused on the chain (according to the EU) of democratic processes ⇒ effectiveness ⇒ better application of politics ⇒ common assessment of the results ⇒ satisfied citizens involved in processes of decision making, reducing of the cost of activities etc.

A Greek Example

The surface of Greece is 130,100 km^2 including the surface of the 3,000 islands that form 20 percent of the total surface area. Two-thirds of this territory is of a mountainous nature and the coastline presents a length of 15,000 km (the longest in Europe), of which 5 percent is of special ecological value. The Greek climate consists of three types that affect specific areas. These types are the Mediterranean type, the Alpine type and the Temperate type. The first, affecting the islands of the Aegean Sea, is characterized by mild, wet winters and warm summers. The second mainly affects the western part of Greece, while the temperate type is encountered in the central and north-eastern parts of Greece. Athens presents a climatic transition with characteristics of both the Mediterranean and Alpine type. It is obvious that this climatic diversity also modifies the water regime of each area.

Greece is a favorite Mediterranean tourist destination because of its natural characteristics. Its rainfall index is higher than that of other Mediterranean countries in the same latitude. This is due to the high mountains near the coastline, especially the Pindos mountain chain which receives most of the rainfall originating from the West or South West.

Despite these particularities, the Mediterranean countries present many common characteristics in terms of water resources. At the same time, they present a range of special characteristics adapted to their general natural and development environment. All the characteristics discussed represent dry or semi-dry climate, limited water resources, and initial agricultural development connected to the particular level of water resources and their economic or social value. Urban and manufacturing development present the same characteristics

International Collaboration on the Mediterranean

The particularities of the Mediterranean basin have been addressed by international organizations dealing with all types of problems including those related to water problems. Examples include the special agreement of UNESCO for the Mediterranean, the special mention of the Mediterranean region in IPCC reports on climate change, etc. The International Association of Hydrogeologists (IAH) participates in special projects concerning the water resources of the Mediterranean basin; an example of their activity is the Report, mentioned in the references, concerning groundwater management and the application of the EU's Frame Directive 2000/60/EU. The "Global Water Partnership – Mediterranean", an International Network of Water-Environment Centers for the Balkans (INWEB) and the International Shared Aquifer Resources Management of UNESCO and IAH are also important collaborative initiatives.

A large number of non-governmental organizations (NGOs) also deal with water in the Mediterranean region. During the 5th World Water Forum held in Istanbul,

15–22 March 2009, the Mediterranean area featured in numerous sessions, and the outcome is expected to be the articulation of many ideas for future regional water strategy. In addition to the special sessions dedicated to water issues in the Union for the Mediterranean, three regional and sub-regional processes covered the full diversity of the region: the Mediterranean as such, the MENA/Arab countries, and the region in and around Turkey. The World Water Council (WWC) and the Euro-Mediterranean Water Information System (EMWIS) agreed to collaborate on improving the collection and dissemination of information and documentation of observation, monitoring and evaluation processes related to the water sector within their respective partnerships. This four-year agreement (2008–2012) includes a particular collaboration of the Water Monitoring Alliance of the WWC based on EMWIS activities related to water information systems and the promotion of a regional water observation mechanism concept and the process developed by the EMWIS as a best practice initiative for possible duplication in other regions. The Assembly of Mediterranean regions and cities (ARLEM) is a new organization which operates on the model of the Euro-Mediterranean Parliamentary Assembly, trying to involve local authorities in projects run by the Union for the Mediterranean. The city of Barcelona (Spain) has been chosen to host the headquarters of the political secretariat of ARLEM. The MEDA Water Programme has given an enormous boost to regional cooperation between different actors in the field of water management. Thanks to this Programme, funded by Europe Aid's Regional Programme, governments have started re-evaluating local water management and have placed it higher on their development agendas. The MEDA Water Programme was put into practice by 69 different organizations in participating countries, 20 NGOs, 23 universities, 16 research institutes and 9 government agencies. The Union of Mediterranean Confederations of Enterprises participated in the Horizon 2020 Capacity Building Sub-Group Meeting which took place in Rome on 9 December 2008. This meeting was organised by the European Commission within the Initiative Horizon 2020 and aimed at drafting a road map for the work programme adopted at Barcelona concerning the three priority areas: industrial emission, municipal waste and urban waste water. For Horizon 2020, the focus should be on measures directly related to the aim of reducing pollution with particular attention to measures linked to pollution reduction projects or monitoring.

References

Andreo, B. and Durán, J. J. 2008. "Theme issue on groundwater in Mediterranean countries." *Environ. Geology*, 54, 443–444.

Annarita, M., Zeng, N., J.Yoon, Artale,V., Navarra, A., Alpert, P. and Z X Li, L. 2008. Mediterranean water cycle changes: transition to drier 21st century conditions in observations and CMIP3 simulations, *Environment Research Letters, 3, 4.*

Boglioti C., Adams R., Avgeropoulos P., Bena C., Foietta P., Giordano A., Guillande R., Stournaras, G., Verdebout, J. 1977. Natural process inducing slope instability and erosion in two distinct geographic regions of the Mediterranean basin in Italy and Greece. *Ann. Géol. Pays Helléniques, v. 37, 797–814.*

COST 620, 2004. Vulnerability and risk mapping for the protection of the carbonate (karst) aquifers. *Final Report*, EUR 20912, 2004.

COST 621, 2004. Groundwater management of coastal karstic aquifers. EUR 20911, 2004.

Deere, D. U., and Patton, F. D. 1971. Slope stability in residual soils, *Fourth Panamerican Conference on Soil Mechanics and Foundation Engineering, American Society of Civil Engineers*, pp. 87–170.

European Commission. 2007. Mediterranean groundwater report. Technical report on groundwater management in the Mediterranean and the Water Framework Directive, *Environment, Technical Report 2007/001.*

IPCC Report on Global Climatic Changes, 2007.

ISSAR A.S. 1993 Once upon a time in Sumeria, *Courrier of UNESCO, Hellenic edition, σελ. 4–8, July.*

Louka, E. (ed.) 2006. Implementing the EU water framework directive towards integrated water management in Europe. *Proceedings of International Conference, Athens, 12 May 2006.*

Margat J., Saad, K. 1984. Les eaux fossiles, *Nature and Natural Ressources, V. XX, No 2, UNESCO.*

UN Commission Economique pour l'Europe, Comité des Problèmes de l'Eau. 1983 Séminaire sur les stratégies et pratiques de protection des eaux souterraines. Athens.

Stournaras, G., Tsimidis, G., Tsoumanis, P., Giannatos, G. and Guillande, R. 1998. Geomorphologic evolution and instability phenomena induced by the lithologic and hydrologic conditions". *Bull. I.A.E.G., V. 57, 65–68.*

Stournaras, G. 1999. Correlating morphometric parameters of Greek Rhone-type deltas. Hydrogeologic and environmental aspects. *Environmental Geology,* 38/1, 53–58.

Stournaras, G, Soulios, G., Bellow. Th., Nikolao, E., Papadopoulos-Orinioti, K.2004. "Karst features as world heritage in Hellas", *Forum on Karst and World Heritage in Europe, Lipica, Slovenia.*

Stournaras, G. 2007. *Water: Environmental Aspect and Evolution. Tziolas* (ed.) *Thessaloniki (in Greek).*

Tissandier, G. 1867. *L'eau. Bibliothèque des Merveilles, Vignettes* par A. De Bar, Glerget, Riou, Jahandier etc., Librairie Hachette et Cie, Paris.

Voropayev, G. 1985. Water and Man, *Courier of UNESCO, Hellenic edition,* March. 4–6.

Wright, E.P. 1985. Groundwater in the third world, *IAH Memoires of the 18th Congress "Hydrogeology in the service of man", V XVIII Keynote papers,* 53–64.

Chapter 10
Water Conservation in Egypt

Nicholas S. Hopkins

Thoughtful Egyptians are beginning to realize that water conservation is essential for the future water supply of Egypt. Virtually all of Egypt's water comes from the Nile River, which is vulnerable to climate change and is governed by the accords and relations with other Nile basin countries (Waterbury 2002). By agreement with the Sudan, Egypt is entitled to 55.5 billion cubic meters of water annually, although it very likely receives more. The per capita amount of renewable fresh water in Egypt is now just under 800 cubic meters, (generally taken as marking the scarcity level), and will decline as population rises. This water scarcity requires mitigation. Roughly 85 percent of Egypt's water is used in agriculture, the rest divided between industry and domestic use. Extreme water shortage is a problem looming on Egypt's horizon. Since the opportunities to increase the available water substantially are limited, careful use of existing water is essential.

Water Conservation

Although any examination of water conservation in Egypt must consider all sectors, any substantial savings must come from the biggest user, agriculture. Savings in agricultural water use depend on high-level decisions (for instance, on the engineering aspects of the irrigation system, or the continuing cultivation of a water-demanding crop like sugar cane). But also much depends on the actions of the myriad of small farmers in the Nile valley. Small farmer practices, including technology and patterns of cooperation, regulate the flow and use of water at the local level, and they interface with the system-wide management of the state.

First I review the practices and behavior of farmers with regard to water use and misuse. A key part of this debate is the issue of getting farmers to pay for irrigation water: whether, why, and how. Use of water for irrigation also implies drainage and the re-use of water, thus we must examine the different ways in which farmers categorize and use water. From a conceptual point of view, there is a trade off between quality and quantity. As the rural population grows, disposal of solid and liquid waste becomes more of a problem; the traditional solution of dumping waste in a canal with running water no longer suffices.

Then I turn to urban dwellers who also have a role to play in water conservation. There are good descriptions of how water is acquired and used in poor urban neighborhoods, where disposal of wastewater is also a problem (Nadim et al., 1980;

Hopkins et al., 2001). Water management (storage and re-use) in the household is a factor. Factories also use a good deal of water.

Most plans for conserving water in Egypt require substantial social participation, and a change in behavior. Yet concerted action is not easy in Egypt, where most forms of social mobilization contravene the emergency laws and disturb the authorities. People may wonder why they should economize on water when their neighbors do not. People tend to feel that water is abundant, because the scarcities are seen as a consequence of the distribution system rather than the availability of water. Only a few know how precarious the overall situation is.

This paper examines current water conservation strategies at the household level in urban and rural areas, and suggests improvements. It also examines the major choices in agriculture, notably efforts to persuade farmers to use water more efficiently – through improved technology, pricing, and the like. The paper concludes with an effort to use present trends to extrapolate water availability and use into the future.

Agriculture

Agriculture in the Nile Valley of Egypt depends completely on the human development of the Nile River. The history of this development goes back millennia, and represents the gradual mastery of the river water for irrigation and other purposes. This history has an engineering dimension and a social dimension. Beginning in the 19th century, the pattern of dikes and canals, designed to have the water reach as much arable land as possible, was gradually systematized and improved under the combined efforts of Egyptian and European engineers. Today, with a growing population and a relatively stable water supply, water conservation is a key issue. This is especially so because the pressure to expand agriculture horizontally into the new lands means the available supply will be stretched further. In fact, a recent Egyptian government report foresees that the amount of water per feddan will be reduced by "over 25 percent" before 2017 as supply is shifted to the new lands (Egypt 2005: 4–27).

Egyptian irrigation is a single system governed by the control mechanism of the Aswan High Dam. This makes water available according to the measures laid down by the Ministry of Water Resources and Irrigation in collaboration with others. The water stored behind the dam flows down the Nile and through the various canals. The canals get smaller and smaller until the water reaches the lowest level of canal, the "*mesqa*". Farmers draw water from the *mesqa* for their fields. The *mesqa* generally has water available about half the time, typically one week on and the next off.

The irrigation system as it exists in Egypt is designed to supply water to the farmer at a level slightly lower than the level of his fields (Hopkins 1987, 1999, 2005; Hvidt 1998). This is known as a "below grade" system, and contrasts with systems where the water flows from the canals into fields by gravity. Thus the

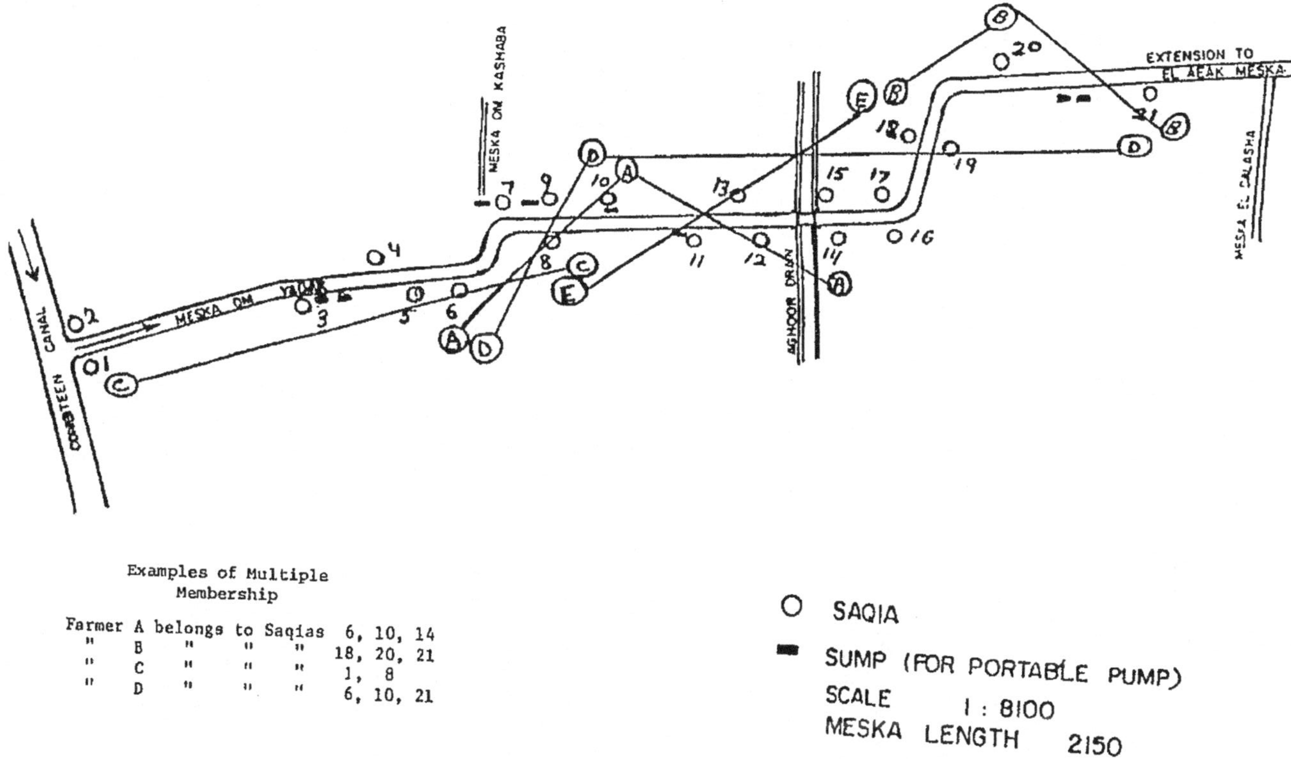

Figure 10.1 Diagram of the *mesqa* "Um Yaddak" (or "Um Aisha") showing the sites of pumps and wheels that lift water to the network of field ditches. The mesqa draws water from the distributory canal. This is characteristic of the Egyptian Delta

farmer must lift water from the *mesqa* to his fields. He may lift as much as he needs from the available water, but he is unlikely to exceed his needs by much because of the extra time and effort required. The water is then conveyed within the fields to the location of the actual irrigation through a network of field ditches. The lifting point defines the boundary between the area under government control (the Ministry of Water Resources and Irrigation, in effect a ministry of irrigation) and the zone managed by the farmers (and linked to the Ministry of Agriculture). The major exception to this system is in the Fayoum where the greater gradient makes gravity flow irrigation possible; in the Fayoum farmers have a time share of a measured flow of water, i.e., a certain quantity (Mehanna et al. 1984, Price 1995a, 1995b).

In Upper Egypt lengthy canals take water from the Nile and convey it downstream at a slightly lower gradient so that it may be used to irrigate fields. In Lower Egypt similar canals branch off from the "barrages" at the apex of the delta. In a few areas between Aswan and Luxor the water is pumped directly from the river into canals. These canals are almost all exposed, and have earthen banks, so conveyance losses to evaporation and seepage are considerable. Once the water reaches the crop, either the entire field is flooded so that water seeps into the soil and hence to the plants, or water is guided between ridges where the crops are rooted. This set of techniques, originating in the old basin system of irrigation, is used effectively by Egyptian farmers to produce some of the highest yields per unit of land in the world. Certainly more water enters the fields than is absorbed by the plants. However, it should be kept in mind that too much water is as deleterious to plants as too little, and Egyptian farmers would not achieve these high yields if they were carelessly over-watering their crops.

On the principle of "clean water in dirty water out," drainage is also critical for the success of Egyptian agriculture. The drainage systems have been improved over the years, from nothing to drainage ditches to the effective "subsurface tile drainage". Water that is drained out is reused, increasing the total amount of available water of the country. Certain styles of irrigation are more subject to water-logging and salinization, and farmers know how to avoid these problems: there needs to be a passage of water through the soil to wash out salt which would otherwise accumulate as the water evaporates.

From the point of view of rural social organization, this description indicates several key points. One is the social organization of the lifting itself, which is generally done by small groups of farmers who must take turns in organizing the lifting, and in organizing the flow through field ditches that often cross the fields of one farmer to reach another. Moreover, this is the point of contact between the canal system managed by the Ministry of Irrigation and the farmers. The Ministry provides water for the *mesqa* according to its guidelines, but the farmers are responsible for maintaining the *mesqa* as an effective transmitter of water. For instance, *mesqa*s must be periodically cleaned and dredged. Higher level canals are fully managed by the Ministry.

Lifting water from the government canals to the private field ditches is typically done either through animal powered water wheels (sing. *saqia*) or through small diesel pumps. The water wheel can have several forms depending on the local technology and history. It belongs to the family of the "Persian water wheels." This system was originally introduced into Egypt about 2,000 years ago and was used to draw water from wells during the recession season. With the introduction of the canal system, the wheels now only lift water from canal to ditch, typically less than one meter. The wheel may have one or several owners. The users, whether owners or renters, provide their own draft animals. The animal turns in a circle to circulate a wheel in the horizontal plane that is geared to a wheel in the vertical plane. This wheel activates a continuous chain of buckets to raise the water, or is attached to a circle of scoops. The users must take turns because the water flow is relatively low, and it must pass through the same network of field ditches. Nowadays most such wheels have been replaced by motorized pumps installed in the same locations and using the same networks of field ditches. The same forms of cooperation at the field level are required. The pumps can be owned by one or several people, and can also be rented. In many areas small movable pumps are used, and transported from place to place. In others, notably in Asyut and Sohag in Upper Egypt, there are large stationary pumps dating from the 1930s when they were used to pump water from wells; they now only raise water from canals. Pumps are employed to save time, while wheels are used when cash flow is a problem, or when only a gentle supply of water is required.

Figure 10.2 Irrigation pump in rural Egypt. Photographer: Nick Hopkins

Among the minor water lifting techniques are the *shaduf*, a leveraged pole with a weight on one end to balance the weight of the water, and the *tambour* or Archimedes screw which uses the principle of the vacuum to lift water. Both are hand-operated, so the cost of use is low, but they only raise water a small distance or in small amounts.

The quantity of water in the conveyance network of canals is certainly an issue, but so is the quality, and they are related as polluted water has limited uses. Water in the system is subject to multiple pollution sources including domestic waste, chemical run-off from agriculture, and industrial waste. The cleaner the water can be kept the more times it can be re-used, thus maximizing the water available to Egypt.

A vexing point is the pricing of water in this system. The idea behind the introduction of the "below grade" system was that if farmers did not have to "pay" for water they would waste it. The effort and cost of lifting water would force farmers to economize. Proponents of the neo-liberal economic system argue that if farmers are to receive market prices for their produce, then they should also pay for the costs of the inputs. This includes the cost of the irrigation water.

Pricing water has run into considerable opposition on the part of many in Egypt. Periodically the Minister of Water Resources and Irrigation has reaffirmed that there is no intention to put a price on Nile water in Egypt (see "*The Egyptian Gazette*," July 3, 2002: 2, also Essam El-Din 2009). A counter-argument has been that although farmers may not pay for the water, they should be expected to pay for the cost of supplying that water. Thus the notion of paying for water is rooted in an ideal of conservation (forcing farmers to economize) and in a (perhaps misguided) theory of the neo-liberal economy. The reluctance of the government to institute such a pricing system is generally argued on the belief that most Egyptian farmers are poor and could not afford the additional cost. Currently farmers do not pay for water. However, the British irrigation engineer Hurst pointed out (1952: 65) that the government provides the water to the farmers without compensation and collects a land tax which also covers the cost of irrigation. Agricultural produce benefits the country's economy in other ways, of course. In all of this, one should look for the larger interests involved, and these include the powerful segment of large farmers as well as the ideology of foreign donors.

What is often not realized by outsiders is that farmers do pay a price in indirect costs for water. They rent pumps, pay for pump fuel, rent draft animals, pay to feed these animals, and must maintain the *mesqa*s and the field ditches. Thus farmer budgets already contain a cost for water. It is just not paid to the government. A farmer who over-irrigates is wasting money.

Farmers are thus careful in their use of water, knowing their crops and their local circumstances. They are aware of the harmfulness of waste, and avoid it. But the system is complicated and small amounts of waste creep in. Better technology and greater attention can minimize this. It is useful to remember that these are real farmers, large and small, trying to make a living for themselves and their families, and for whom irrigation is a taken-for-granted tool. They are not perfect consumers.

Changes in Water Use

Are there changes that could ensure the continued high level of productivity of land in Egypt while at the same time conserving water so that more can be used elsewhere? One obvious change would be to modify the crop mix of Egyptian agriculture to reduce the importance of such water-demanding crops as sugar and rice. For political reasons this option may not seem right, but it should remain at least on a side table during the discussions. The reduction in either crop would have direct effects on the linked industrial and commercial structures, such as sugar factories and rice milling and marketing.

Technological changes are also available, for instance to change the system in the old lands of Egypt away from "flood" irrigation and towards a combination of sprinkler and drip irrigation, as often used in the new lands. This would enable a more efficient use of water, but quite a different style of agriculture. The major drawback here is the high cost of the capital investment, especially for the small farmers who are the overwhelming majority of Egyptian farmers. The need for water pressure means some additional operating costs. Some small-scale experiments have been carried out with the help of foreign aid donors, but so far these experiments have not spread. There may also be some environmental consequences; the experience of Jordan's East Ghor project has shown that repeated use of lower-water techniques can lead to salinization (Elmusa 1994); this should be examined. Another change would be to focus on minimizing conveyance losses (evaporation and seepage) by re-engineering open canal systems into covered piped systems. Each of these changes requires careful study; since the present system "works" one should be cautious not to "fix" it into something less effective. Sooner or later, population and economic pressures will require moves in this direction. However, it is important that planners keep not only engineering goals but also social goals such as equity in mind (Ayeb 2002).

Water Quality

Water flows through Egypt typically in open canals, with a parallel set of drains for used water. Dumping waste in a convenient canal that can carry it away is a constant temptation. Some feel that flowing water purifies itself (which is possible but not within the distances involved). The flowing water that removes waste downstream also brings some in from upstream, so the gain may be illusory. Excessive dumping diminishes the water quality, and thus harms the downstream users. Since polluted water is used for irrigation, contaminating plants, the damage can be more extensive. Research carried out in Minufiyya in the mid-1990s showed that all the water sources available to villagers were contaminated – the canals, the shallow wells, and the piped water supply from artesian wells, although for different reasons. Not only chemical pollution, but water-borne diseases such as schistosomiasis are endemic (Hopkins et al. 2001:42–43, see also El Katsha and Watts 2002).

The Egyptian parliament has passed laws forbidding such dumping and imposing fines. Most people are aware of such laws, for instance, Law 48 of 1982. In our sample in the late 1990s, 71 percent admitted to knowledge of this law (Hopkins et al. 2001:104). They understand that such dumping is the major cause for poor water quality. Nonetheless they continue dumping (and tolerating other forms of pollution, such as agriculture run-off) in part because alternatives are not available. Dumping can entail a fine of LE1000 but the law is not enforced. We have no record of people chiding their upstream neighbors for fouling the canals (Hopkins et al. 2001:100–104). Mehanna et al. (1984:57) report that the dumping in the canals adds a cost for the farmers as they must hire machinery to clean out the hazardous metal and broken glass illegally dumped in the canal bed.

Conserving Water in Agriculture

Actual water loss in the Egyptian irrigation system is about equally divided between loss from the canals and other water works and loss (inefficiency) from the ways in which farmers use water in the fields (Radwan 1997, 1998). The first problem can be solved by better, but very costly, engineering (a switch from canals to pipes, for instance). A lot of discussion has revolved around the second problem. The point of the discussion has been to devise ways to persuade, or teach, the farmers how to use water more efficiently. Doubtless there are tips that can be passed on. However, there are probably limits to how much water can be saved in this way, given the present technology. Much more could be saved by moving away from water demanding crops like sugar cane and rice, or by introducing sprinkler or drip irrigation in the old lands. (These systems predominate in the new lands.) These methods depend on a government policy decision and have considerable costs for the government. It would be unrealistic to expect Egypt's small farmers to bear the general investment cost of sprinkler or drip irrigation. A minority probably could link themselves to a larger system if one were available.

From a certain point of view, then, placing the emphasis for water loss on farmers amounts to a form of blaming the victim – they are not responsible for many of the losses and are not in a position to make the massive investments needed to move to a more conservationist level. Under the present system the farmers are fairly efficient, and with the best skills could probably save about 10–20 percent of the water that comes their way – very roughly 5 percent of the water in Egypt: not negligible, but not earth-shattering either.

However before one decides that applying 5 percent less water also results in 5 percent more water available system-wide one must answer the question of what the fate of the over-irrigation was. It has been noted that when irrigation systems become more efficient, downstream users who irrigated with the excess water suddenly suffer water shortages

Domestic and Industrial Use of Water

There has not been much research about the potential water saving of reforms in the Egyptian industrial sector. This wastewater is often so heavily polluted as to be a danger to the human population. Currently, the pollution passes down drains until it reaches the sea or evaporates. Sometimes these drains are veritable corridors of disease. Some waste water passes through government water purification systems, but the obvious improvement here is purification treatment in the factory before the water is released into the drainage system.

Domestic use of water amounts to about half the 15 percent of water that is not used in agriculture. Historically, people simply took water from the river or from the canals, and in many rural areas there was considerable scarcity of water when the canals were low. Since the revolution of 1952, successive governments have made an effort to provide piped water not only in the towns and cities but also in the rural areas. Today almost all settlements are provided with piped water. However, quality issues remain. The Minufiyya village study mentioned above showed that the piped water was of dubious quality because of unclean storage tanks and leaky pipes. Most people relied on their own shallow wells for drinking water, but this too was of poor quality, though it passed the "tea test" by making tastier tea and was considered good. For some purposes (washing clothes and dishes) people used water from the canals, frequently by taking their washing to the canal. Even where water was relatively plentiful, people sometimes had to ration their use of it because of the lack of ways to dispose of the waste water – villages are only beginning now to have sewage systems, and they are often built by local people so the quality varies considerably. These locally-built sewage systems are not connected to a master system, but simply dump the untreated sewage away from the village downstream towards the next village. In the absence of even this basic sewage system, village women manage their water carefully, often reusing it within the household so as to minimize waste (el Katsha 1989).

Somewhat the same pattern is true for urban areas that lack a sewage system or in some cases, piped water at all. A survey of areas of Cairo without piped water in the late 1970s showed that people were extremely conscious of the problem of disposing of waste water. They did not want to bring more water into their neighborhood than they could use or discard for fear of creating standing pools of excess water (Nadim et al. 1980). Cairo and other urban areas are growing so rapidly that provision of services, especially water and sewage, does not keep up. Madinat Nasr, Helwan, and other Cairo neighborhoods suffer water problems in late summer, mostly because population growth has outstripped the capacity of the supply system. As the city expands, more water must be pumped to higher elevations and that can be a source of breakdown as well (Hamdi 2003; Leila 2008).

Where water delivery and sewage systems are in place, water usage patterns approximate western models. Urban Egyptians are essentially apartment dwellers; single-family homes are the exception. Users are billed for water proportional to the level of use. In some cases it is not the user but the building that is billed, on the assumption that the owner of the building pays the bill and recovers his expense through the rent. In older buildings with rent control, the rent no longer covers the water bill or much of anything else. Then the owners leave responsibility for the water bill to the occupants who must organize among themselves. In areas without piped water, many residents rely on water carriers to bring them water, and usually pay more than the bill for piped water would have been.

Again in the urban areas there are conveyance losses and losses due to leaky taps in apartments. There are also some visible wasteful practices, such as car washing and spraying water through hoses to clean sidewalks and streets. Water is also used with abundance to wash inside floors. How much water is lost through these various defects and poor practices is unknown, but is probably substantial. Some of these losses could be corrected by better engineering, and some would respond to more conservationist habits (closing taps tightly, being sparing in the use of water for cleaning floors, sidewalks, cars, etc.).

Solutions

The geographer Habib Ayeb has noted that "Water management ... requires a very long-term policy that combines the rationalization of demand and modes of consumption with a large-scale introduction of new technologies limiting water waste via the modernization of the entire hydraulic system" (Ayeb 2002: 80). He then proceeds to stress the importance of local patterns of organization and self-management in water conservation.

Two principal solutions leading to water conservation are on the one hand an effort to mobilize people so they internalize a conservationist ethos and are careful in their use of water, and on the other, price mechanisms. One can use pricing to encourage limits on water use among really big users (industrial, agricultural, or othe)r, but one has to be careful not to raise the price for the poor bulk of the population who may simply react by reducing their water consumption below the levels desirable for health. The argument so far has been that the discretionary use of water that might be affected by either or both of these mechanisms is relatively small, perhaps on the order of 5 percent for the country as a whole. Five percent more water is not to be disregarded; at the same time, with the population growth rate above 2 percent the extra water would soon be needed for the additional population.

Much greater savings of water might come from various engineering solutions, for instance, improving the distribution of water through pipes in the urban areas (in 2003, Hamdi was quoted the figure of a 25 percent loss, five times the norm; some estimates are higher), and eliminating the conveyance losses due to evaporation and seepage in the irrigation sector. A shift from flood irrigation to sprinkler and drip irrigation would also reduce water demand, but the engineers would have to be careful to avoid salinization of the soil. Moreover, if the use of sprinkler and drip irrigation in the old lands reduces the amount of water eventually returned to the Nile, the system wide water savings could be small. More field research is needed on the impact of water saving techniques.

A solution not yet publicly discussed is water rationing, presumably by cutting off water supply to canals or urban neighborhoods on a rotating basis. The water supply in canals already rotates, as we have seen, but the rotation could be extended. In the urban areas, cut-offs to neighborhoods act as a form of water rationing whether intended or not. This form of water rationing is known from elsewhere in the Arab World, notably Jordan.

Given the inexorable growth of population, if the water supply remains largely constant, some form of water conservation is going to be forced on Egypt sooner or later. From a social science point of view, the issue is whether this will happen through a mobilization of the population (organization of demand), or through a control of the distribution system that will impose limits on the ability of people to access water (restrict supply).

References

Ayeb, H. 2002. Hydraulic politics: The Nile and Egypt's water use: a crisis for the 21st century? in *Counter-Revolution in Egypt's Countryside: Land and Farmers in the Era of Economic Reform*, edited by Ray Bush, London: Zed. 76–100.

Egypt 2005. *Water for the Future, National Water Resources Plan, 2017*. Cairo, Ministry of Water Resources and Irrigation.

El Katsha, S. et al. 1989. Women, water, and sanitation: household water use in two Egyptian villages. *Cairo Papers in Social Science* .12(2).

El Katsha, S. and Watts, S. 2002. *Gender, Behavior, and Health: Schistosomiasis Transmission and Control in Rural Egypt*. Cairo: AUC Press.

Elmusa, S.S. 1994. *A Harvest of Technology: The Super-Green Revolution in the Jordan Valley*. (Washington: Georgetown University, Georgetown Studies on the Modern Arab World).

Essam El-Din, G. 2009. Minor shake-up, *Al-Ahram Weekly*, March 19, 3.

Hamdi, O. 2003. Quartiers en panne sèche, *Al Ahram Hebdo*, Sept. 3, 2003:7.

Hopkins, N.S. 1987. Mechanized irrigation in upper Egypt: the role of technology and the state in agriculture, in *Comparative Farming System,* edited by B.L. Turner II and Stephen B. Brush. New York and London: Guilford.

— 1999. Irrigation in contemporary Egypt, in *Agriculture in Egypt from Pharaonic to Modern Times,* (ed.) Alan K. Bowman and Eugene Rogan, Proceedings of the British Academy 96, 367–385.

— 2005. The rule of law and the rule of water in 'Le Shaykh et le Procureur', *Egypte/Monde Arabe*, edited by Baudouin Dupret and François Burgat, 3e série, (1).

Hopkins, N.S., Sohair Mehanna, and Salah el Haggar. 2001. *People and Pollution: Cultural Constructions and Social Action in Egypt*. Cairo: AUC Press.

Hurst, H.E. 1952. *The Nile*. London: Constable.

Hvidt, M. 1998. *Water, Technology and Development: Upgrading Egypt's Irrigation System*. London: Tauris.

Leila, R. 2008. No flow: fresh potable water is becoming a scarce commodity" in *Al-Ahram Weekly*, 9, 4.

Mehanna, S. et al. 1984. Irrigation and society in rural Egypt, in Nadim, A. et al. *Cairo Papers in Social Science* 7(4). Living without water, *Cairo Papers in Social Science* 3(3), 1–132.

Price, David. H. 1995a. Water theft in Egypt's Fayoum oasis: emics, etics, and the illegal in, *Science, Materialism and the Study of Culture*, edited by Martin F.Murphy and Maxine L.Margolis, Gainesville: University Press of Florida, 96–110.

Price, David. H. 1995b. The cultural effects of conveyance loss in gravity-fed irrigation systems, *Ethnology* 34(4),273–291.

Radwan, Lutfi. S. 1997. Farmer responses to inefficiencies in the supply and distribution of irrigation requirements in Delta Egypt, in *The Geographical Journal.* 163(1),78–92.

Radwan, Lutfi. 1998. Water management in the Egyptian Delta: problems of wastage and inefficiency, in *The Geographical Journal* 164(2),129–138.

Waterbury, J. 2002. *The Nile Basin*. Yale University Press.

Conclusion

The Mediterranean may be, as I suggest in the initial chapter of this book, a region of imagined unity, but one thing all the authors in this volume reiterate is that the countries that border the Mediterranean Sea are confronted with a severe crisis of fresh water resources. Each country's approach to the crisis, even if that nation is a member of the EU, is different, and conditioned not only by the particularities of its environment, but by cultural attitudes as much as by local political and economic conditions. Spain and Greece, two EU counties that face severe water problems, might be expected to have common approaches to the water crisis, yet, as Gaspar Mairal points out, public policy has deep cultural roots, and in Spain the issue of water acquired Biblical dimensions in the 19th century in strong contrast to Greece, where water was never a central issue of national policy.

In both Spain and Greece, democracy is a relative novelty. In Spain, with its history of grandiose water policy, water soon became a party issue, "more politics than policy". In response to national dissatisfaction with water policies and escalating tensions caused by water shortages, a broad-based social movement has developed in Spain around water. "The New Water Culture", involving students, academics, politicians and agriculturalists in discussions of water policy, is unique in the Mediterranean and no doubt owes its success to the historical importance of water at the national level. In Greece, issues of national security, historical patrimony and political corruption have taken precedence over environmental issues, and water policy has been largely ignored.

The 2000 EU Water Framework, which should ideally be the basis for water policy in European countries, does not overcome national differences in water policy, nor in the implementation of its regulations. In his chapter, Eriberto Eulisse demonstrates how commercial interests in Italy have undermined both the traditional respect for water in the country, and environmentally-sound policy. A prime example is the bottled water industry, which has persuaded Italians that water can only be safe if it comes in a labeled container. The consequences of the powerful lobby of bottled water companies for Italy is the degradation of the environment and the destruction of a fundamental cultural relationship between the inhabitants of the Italian peninsula and their sources of fresh water.

Despite differences in their natural water resources and national water policies, the countries of the northern Mediterranean littoral are obviously more fortunate than those of the south or east. In the Middle East and North Africa the issue of water has already passed the stage of seasonal scarcity and become calamitous. In a chapter on the relationship of water to international law, Keith Porter addresses the much-discussed issue of water as a human right. This leads to the question of how such a right might be enforced. Porter concludes that however difficult it may be, the right to water must be seen as a moral obligation, necessary for human welfare and economic development. Governments who do not have the means to provide their citizens with adequate water must be helped by those who do.

Insufficient water resources are not only a cause of human suffering and lack of development, but may lead to conflict between nations and territories. There has been much written in recent years about the danger of "water wars" in the 21st century. Once regarded as the most fortunate of the Middle Eastern countries in terms of its water resources, Lebanon is suffering from water problems that are directly related to its 20 year history of civil war and international conflict. Not only have bombs damaged the infrastructure and polluted the water supply of Beirut and other Lebanese cities, but a state of continual national crisis has destroyed the civil framework necessary for sound water planning.

If the quantity and quality of water in the Mediterranean countries has already reached crisis level, and increasing population, particularly in urban areas, continues to put pressure on the available supply of water, what can be done to reverse or ameliorate the problem? Tammo Steenhuis stresses the importance of understanding the water balance within a watershed basin. Once this has been done, a water "budget" can be drawn up that considers the total amount of water available and how that water may be equitably distributed. Zeyad Makramreh echoes the need for careful management of water resources in the eastern Mediterranean. He points out that cooperation between Middle Eastern countries is the only possible strategy for dealing with the region's water problems. Technical solutions, such as recycling water for agricultural purposes and desalinization, are both only partial fixes.

On Cyprus, despite the partition of the island into two independent entities, water management is increasingly recognized as a bi-communal issue that must be addressed by Greek and Turkish Cypriots working together. Despite its arid climate Cyprus was once able to store enough water in its dams to supply the needs of the entire population through the dry summer months, but increasingly high summer temperatures and lack of rainfall have combined to lower the dams to a dangerous level. One solution to the crisis on the island, according to Symeon Christodoulou, is to reduce the wastage of water in urban pipes, something that is already being implemented in parts of Cyprus. Like Cyprus, Greece has been severely affected by climate change. Despite its specific local problems, geologist George Stournaras stresses the importance of the collaborative efforts made in recent years by international organizations that address the region as a whole.

Improving the efficiency of urban water distribution is one of the solutions Nick Hopkins proposes for the looming crisis facing Egypt. He notes that rapid population growth will force some form of water conservation on the Egyptians, although it is not easy to predict how this will be achieved. Popular mobilization is one possibility; another is some form of government control that will either force larger users to pay for water or restrict their access to it.

What, if anything, can we conclude from this collection of essays written by scholars from a wide variety of disciplines, about the water crisis in the Mediterranean?

The Mediterranean is the still the most popular tourist destination in the world, beloved not only for its sun and sea, but for its cultural and historical associations. Are we bound to lose a "paradise" that has drawn people to these shores for thousands of years? Can we avoid a crisis that will cripple agriculture, destroy wetlands, rivers and streams, and cause provinces and countries to fight one another over dwindling water resources? As the authors of this volume agree, there is no easy answer. Neither international regulations nor laws declaring water a human right will stop people dying of thirst and water-related diseases. Nor will they stop developers from building massive hotels and golf-courses in ecologically fragile areas. Corruption and economic necessity make nonsense of regulations in many of the countries in the region. There are technical solutions that can help conserve water by making water distribution in urban and rural areas more efficient, or by re-using water and desalinating it, but these are expensive solutions, dependent on careful, coordinated supervision that is not available in every Mediterranean country.

If there is any hope for rescuing paradise, it must come from more than one source, even the unexpected. In some countries, as Eriberto Eulisse suggests, it may involve going backwards rather than forwards, by reminding the population of Italy, for example, that water was once a sacred substance, not a commodity to bottle and sell for profit. Cooperation is an obvious necessity for any progress in the region. The citizens of Egypt, Greece, Spain and Morocco must recognize the need to confront the water crisis together. Children must also be taught the need to conserve water at school. They will soon be facing a crisis of even greater severity. We who care about the crisis of water in the Mediterranean must recognize that none of us has the solution to the problem. In an increasingly arid Mediterranean, government officials and local administrators, scientists, farmers, artists and citizens must work with one another, devising imaginative solutions, pooling their knowledge, as we have in this book, to rescue paradise before it is truly lost.

Index

Numbers in italics refer to figures/illustrations.
Numbers followed by "t" refer to tables.
Numbers followed by "n" refer to notes.

For Product Safety Concerns and Information please contact our EU representative GPSR@taylorandfrancis.com Taylor & Francis Verlag GmbH, Kaufingerstraße 24, 80331 München, Germany

Printed and bound by CPI Group (UK) Ltd, Croydon, CR0 4YY
14/09/2026
02216223-0018